Bioinformatics and Computational Biology

Basant K. Tiwary

Bioinformatics and Computational Biology

A Primer for Biologists

Springer

Basant K. Tiwary
Department of Bioinformatics, School of Life Sciences
Pondicherry University
Puducherry, India

ISBN 978-981-16-4240-1 ISBN 978-981-16-4241-8 (eBook)
https://doi.org/10.1007/978-981-16-4241-8

This Springer imprint is published by the registered company Springer Nature Singapore Pte Ltd.
The registered company address is: 152 Beach Road, #21-01/04 Gateway East, Singapore 189721, Singapore

Preface

The goal of writing this book is to explain the theoretical foundation and mathematical concepts of bioinformatics and computational biology to a biologist who is new to this subject. This book is aimed at undergraduate and graduate students of any biological discipline like life sciences, genetics, medicine, agriculture, animal husbandry and bioengineering in addition to the students of bioinformatics and computational biology. Like any other technical subject, bioinformatics has its own language and terms. All terms are explained in simple language avoiding mathematical jargon. This textbook starts with basic theoretical concepts in bioinformatics building in a stepwise fashion leading to an optimum level where the application of bioinformatics in medicine, agriculture, animal husbandry and bioengineering is apparent. This work examines the underlying principles and methods in bioinformatics applied to answer real biological problems. The last four chapters are completely applied in nature although theoretical concepts discussed in earlier eight chapters are a prerequisite for understanding the applications. The exercises and multiple-choice questions at the end of each chapter will reinforce the understanding of the concepts discussed in the text. The worked-out answers to exercises in each chapter will provide a thorough understanding of the problems in bioinformatics and their stepwise solutions. Some useful references are added at the end of each chapter for further reading.

I am indebted to all my teachers who have shaped my understanding of interdisciplinary science. I thank my Ph.D. advisor, Prof. Arun K. Ray, who has given me the freedom to explore the interdisciplinary and experimental biological research at Bose Institute. I express my heartfelt gratitude to Prof. Wen-Hsiung Li, James Watson Professor at the University of Chicago to initiate me in the field of computational genomics during my stay in his lab as a visiting faculty. I am also grateful to all research workers across the world in the field of bioinformatics and computational biology for building the knowledge base of the book. I thank all my colleagues and students at the Department of Bioinformatics, Pondicherry University, for creating a wonderful teaching-learning environment. Special thanks to Dr. R. Krishna for reading and commenting on the chapter on structural bioinformatics. I also thank the team at *Springer Nature* for making this academic endeavour successful.

Puducherry, India　　　　Basant K. Tiwary

Contents

1 Introduction to Bioinformatics and Computational Biology

Learning Objectives

You will be able to understand the following after reading this chapter:

- Definition of bioinformatics and computational biology.
- Connection between bioinformatics and other biological and physical sciences.
- Difference between computational thinking and computational programming.
- Evolution of bioinformatics as an interdisciplinary subject.

1.1 Introduction

Biology has undergone a sea change in the last two decades and transformed itself into a true quantitative science akin to mathematics, physics and statistics. Although biology was fortunate enough to be enriched by great mathematicians like G.H. Hardy, R.A. Fisher, J.B.S. Haldane and Sewall Wright during the last century, it continued to remain largely a descriptive natural subject with less emphasis on quantitative methods until the end of last century. The advent of fast microcomputers and advanced molecular biology techniques such as PCR, microarray and next-generation sequencing has led to the emergence of high-throughput data in biology. Surprisingly, the generation of genomic data has surpassed astronomical data in terms of volume and became the most abundant available big data today. These drastic developments in generation of biological data has necessitated a parallel development of computational methods to analyse the data in a meaningful way and ultimately led to the emergence of a new discipline known as bioinformatics. Although biology and bioinformatics are currently being treated as separate

B. K. Tiwary, *Bioinformatics and Computational Biology*,
https://doi.org/10.1007/978-981-16-4241-8_1

disciplines, the time is not far away when a thin line of division between these two disciplines will gradually disappear in favour of a more holistic biology.

1.2 Definitions of Bioinformatics and Computational Biology

Bioinformatics and computational biology has been defined as distinct subjects initially albeit a significant overlap at the interface of two disciplines. The NIH Biomedical Information Science and Technology Initiative Consortium in the year 2000 proposed the working definitions of Bioinformatics and Computational Biology. Bioinformatics was defined as the research, development or application of computational tools and approaches for expanding the use of biological, medical, behavioural or health data, including those to acquire, store, organize, archive, analyse or visualize such data. On the other hand, computational biology was defined as the development and application of data-analytical and theoretical methods, mathematical modelling and computational simulation techniques to the study of biological, behavioural and social systems. In simple words, bioinformatics is the biological application of information technology with focus on data storage, whereas computational biology is an application of information technology in understanding biology with more emphasis on analytical algorithms. In other words, computational biology is a computational approach to analyse and understand biological processes. However, under current scenario, many areas in these two disciplines are overlapping, henceforth, we will use the term interchangeably in the rest of the book.

1.3 Interdisciplinary Science

The development of bioinformatics occurred at the interface of various scientific disciplines such as biology, computer science, mathematics and statistics, physics and chemistry (Fig. 1.1). The key areas in biology required for understanding bioinformatics are biophysics, biochemistry, cell and molecular biology, genomics and evolutionary biology. A good understanding of some fundamental concepts in computer science such as programming, database, data structures, machine learning and artificial intelligence are prerequisite for learning bioinformatics. In addition, biostatistics, probability theory, linear algebra, discrete mathematics, differential equations, Bayesian statistics and calculus have contributed significantly in the development of bioinformatics. Bioinformatics also draws ideas from fundamental concepts in physics and chemistry for finding solutions to a computational problem.

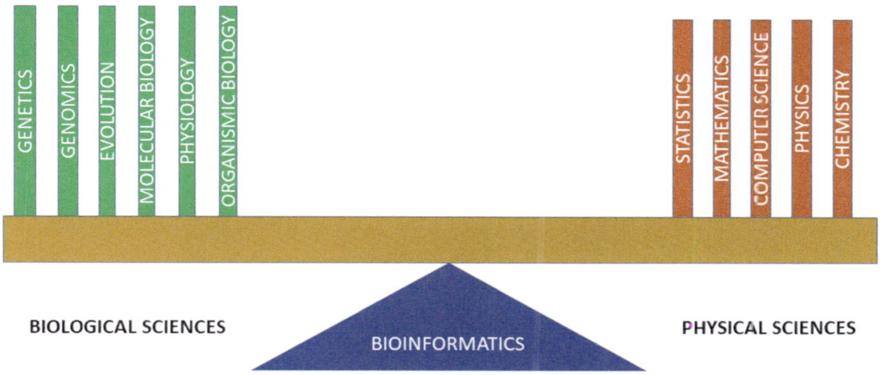

Fig. 1.1 Bioinformatics as an interdisciplinary subject at the interface of biological and physical sciences

1.4 Computational Thinking

Can we develop expertise in bioinformatics with minimal knowledge in computer programming? Surprisingly the answer is yes! The simple logic behind this paradoxical answer is that computational thinking is different from computer programming. Computational thinking is a logical thought process encompassing formulation of a complex problem and its subsequent possible computational solutions in four steps (Fig. 1.2). First, a complex problem is broken down into smaller parts (decomposition), followed by finding some similarities among smaller parts (pattern recognition), and focussing on key information (signals) only avoiding concomitant unnecessary noise (abstraction) within and among the parts. The final step of computational thinking is the development of solution to a problem in stepwise manner (algorithms). On the other hand, computer programming is merely one aspect of software development having four successive steps: analysis, design, implementation and evaluation. The first and last step of software development is only based on computation thinking and does not need any knowledge in programming. Thus, a computational biologist should be able to ask a biological question and find its solution through algorithmic thinking. He should also be capable of writing a new computational program or customize an existing program to his biological analysis.

1.5 Bioinformatics Skills

A bioinformatics scientist is expected to have an array of computational skills. These include ability to manage, store and analyse large biological datasets using available algorithms and software. In addition, he should be an expert in mathematical and

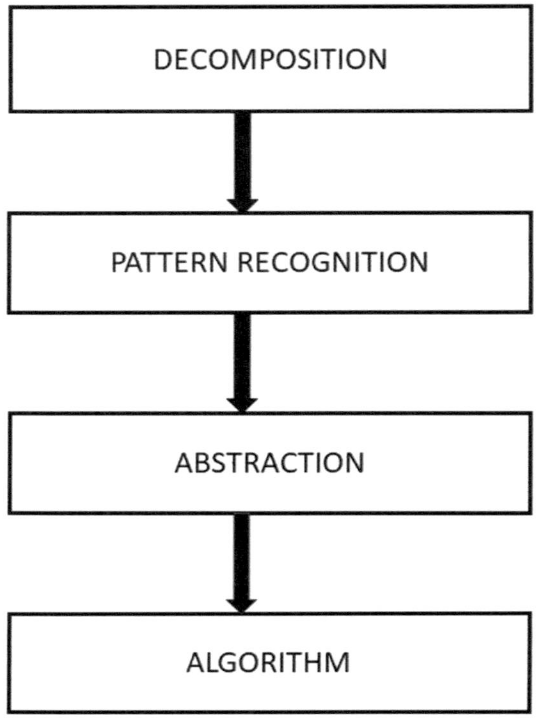

Fig. 1.2 Four essential steps in computational thinking

statistical modelling with strong foundation in the probability theory, graph theory, descriptive and inferential statistics, differential equation and statistical programming in the R language and environment. He is also expected to regularly maintain and upgrade systems and servers in his lab. The knowledge of one scripting language like R or Python will have an additional advantage. A good understanding in the core biological subjects such as genetics, genomics, biochemistry, molecular biology and evolution are prerequisite for a bioinformatics scientist. A continuous learning of advanced technologies like next-generation sequencing and mass spectrometry is also required. Above all, a bioinformatics scientist should be highly motivated and dedicated person and have a teamwork spirit with excellent analytical ability.

1.6 Historical Perspective

Bioinformatics has undergone a magnificent journey with an extraordinary pace in the last seven decades. The first biological problem solved by bioinformatics approach was to decipher the protein primary structure using Edman peptide sequencing data. Margaret Dayhoff, the mother and father of bioinformatics, developed a FORTRAN program called COMPROTEIN between 1958 and 1962 for finding this solution. Emile Zuckerkandl and Linus Pauling started evolutionary analysis of several protein biomolecules including haemoglobin in 1963. This evolutionary analysis conclusively demonstrated two facts: orthologous proteins in vertebrates have a high degree of similarity in sequence over a long evolutionary period and the degree of differences among orthologous sequences from different species were proportional to evolutionary distances between species. Zuckerkandl and Pauling concluded that all these orthologous sequences were evolved from a single common ancestor based on the above-mentioned facts. This novel idea in evolutionary biology further paved the way for prediction of an ancestral sequence from the available sequences of extant species.

Both molecular biology and computer science developed at fast pace albeit in parallel from 1980 to 1990. The emergence of new molecular techniques like gene cloning and PCR revolutionized the biological and clinical research. Concomitantly, microcomputers were invented in 1977. Needleman and Wunsch developed an algorithm for the first time for aligning protein sequences in 1970. This algorithm further led to emergence of multiple sequence alignment (MSA) in early 1980s although this algorithm was useful but impractical to apply due to their high computational running time. Consequently, Feng and Doolittle in 1987 developed another approach to MSA known as progressive sequence alignment based on a guide tree. Most popular program for multiple sequence alignment CLUSTAL was derived from this algorithm. First probabilistic model of amino acid substitution was developed by Dayhoff, Schwartz and Orcutt in 1978 in form of point accepted mutations (PAMs). With the introduction of this PAM matrix, amino acid substitutions became a popular metrics for measuring evolutionary changes in a sequence. The development of DNA sequencing technologies has paved the way for whole-genome sequencing of many organisms.

After completion of human genome project, massive worldwide efforts started to sequence various animal and plant genomes. DNA sequencing of genomes has become much cheaper and faster with time creating huge amount of genome sequences as a challenging big data for computational biologists. A new software, Staden was developed to analyse the Sanger sequencing reads. The first phylogenetic tree was reconstructed using protein sequences based on the least number of amino acid changes (maximum parsimony method). But DNA sequences carry more information than the protein sequences in terms of synonymous mutations which does not manifest in a protein sequence. Statistically more robust method known as maximum likelihood method was developed for inferring phylogenetic tree by Felsenstein using DNA sequences. This method finds a tree which has maximum probability of evolving the observed data. Bayesian approach to molecular phylogeny gained momentum in the 1990s and is a very popular and statistically most robust method of phylogenetics.

The first software developed for sequence analysis was the GCG software suite in 1984 to manipulate sequences on small-scale mainframe computer. DNASTAR, alternative software suite for personal computer was developed in the same year. Three international databases, namely European Molecular Biology Laboratory (EMBL) database, GenBank and DNA Data Bank of Japan (DDBJ) were developed independently in Europe, the USA and Japan, respectively. Finally, these three databases were integrated in 1986–1987 and are still existing under the umbrella of the International Nucleotide Sequence Database Collaboration. Many scripting languages were developed in the 1980s for application in bioinformatics. These scripting languages are interpreted whenever launched and do not require compilation from C code. Perl (Practical Extraction and Reporting Language) was developed in 1987 and soon became a popular language to manipulate biological sequences. Due its flexibility, it became very popular among bioinformaticians and improved further in form of BioPerl in 1996. Today, R and Python are two major programming languages for bioinformaticians.

The first genome of *H. influenzae* was sequenced by Craig Venter and his team in 1995. But human genome project was completed with combined private and public efforts using Sanger sequencing technology in 2003. At present, genome sequencing is much cheaper after the advent of second-generation or next-generation sequencing technologies. The 454 pyrosequencing technology was the first next-generation sequencing platform. But Illumina platforms like HiSeq soon became a popular sequencing platform for high-throughput sequencing. This has led to development of many alignment software for assembly of short reads generated by various sequencing platforms. Today, high-performance computing facility is being provided by various public and private agencies.

Now, modern biology is moving from reductionist phase which focuses on a single gene or a protein to a more holistic approach of systems biology. A mathematical model representing a biological process can be generated from genome or transcriptome or proteome or metabolome data. This endless journey of bioinformatics from sequence analysis to systems biology will continue with a much faster pace in near future. Computational biology is likely to be an integral part of agriculture and medicine worldwide soon for a better food production and healthcare.

1.7 Goals of the Book

This book is intended for biologists who have a basic understanding of mathematical concepts. Bioinformatics is a diverse discipline with a potential to solve real problems in biological, medical, agricultural, veterinary science and bioengineering. This book is a humble attempt to communicate the concepts in bioinformatics to experimental biologists with minimum emphasis on the underlying mathematical principles. The mathematical descriptions are simplified in favour of a better understanding of biological processes. The first part of this book deals with fundamental concepts in biological databases, statistical computing in R, sequence alignment, structural bioinformatics and molecular evolution. A good concept in biological database and sequence alignment are essential for understanding any advanced field in bioinformatics. R language and environment has become the lingua franca of bioinformatics and computational biology with largest collection of statistical packages available for biological data analysis and visualization. I will discuss the molecular structure and dynamics of macromolecules in connection to their functions in a chapter on structural bioinformatics. This will be followed by understanding the molecular basis of genome evolution in the next chapter on molecular evolution with various methods of inferring evolutionary relationship from DNA and protein sequences. The second part of this book will focus on two advanced areas of bioinformatics, namely the next-generation sequencing and systems biology. The third part of this book deals with application of bioinformatics in clinical, agricultural, farm animal informatics and bioengineering research.

All chapters have exercises and multiple choice questions for self-assessment test by the reader. Two exercises are included in each chapter to illustrate the application of bioinformatics in various branches of basic and applied biological sciences. I wish that this book will be a useful companion to not only experimental biologists and other colleagues from different branches of physical and mathematical sciences but to a computational biologist as well.

1.8 Multiple Choice Questions

1. Name a biological discipline which is NOT an integral part of bioinformatics:
   (a) Molecular Biology
   (b) Evolutionary Biology
   (c) Environmental Biology
   (d) Biochemistry
2. Who is known as the mother and father of bioinformatics?
   (a) Margaret Dayhoff
   (b) Francis Crick
   (c) Sydney Brenner
   (d) Motoo Kimura
3. Who proposed for the first time that orthologous sequences were derived from a common ancestor?
   (a) Zuckerkandl and Pauling
   (b) Kimura and Nei
   (c) Nei and Mayer
   (d) Mayr and Goldman
4. Who developed progressive multiple sequence alignment based on a guide tree?
   (a) Feng and Doolittle
   (b) Smith and Waterman
   (c) Wunsch and Needleman
   (d) Smith and Needleman
5. The first method used for reconstruction of a phylogenetic tree from protein sequences is:
   (a) Maximum parsimony
   (b) Maximum likelihood
   (c) Neighbour joining
   (d) UPGMA

6. Maximum likelihood method for reconstruction of a phylogenetic tree was developed by:
   (a) Felsenstein
   (b) Kimura
   (c) Nei
   (d) Goldman
7. First software for sequence analysis is:
   (a) GCG
   (b) MEGA
   (c) BIOEDIT
   (d) PHYLIP
8. Which of the following is NOT a scripting language?
   (a) R language
   (b) Perl
   (c) Python
   (d) HTML
9. Which of the following is NOT as part of International nucleotide sequence collaboration?
   (a) GenBank
   (b) EMBL
   (c) DDBJ
   (d) UNIPROT
10. Name the microorganism whose genome was sequenced for the first time:
    (a) *H. influenzae*
    (b) *E. coli*
    (c) *M. tuberculosis*
    (d) *Mycobacterium pullorum*

Answer: 1. c 2. a 3. a 4. a 5.a 6. a 7. a 8. d 9. d 10. a

Summary

- Bioinformatics and computational biology are integral parts of holistic biology.
- Bioinformatics deals with development of computational tools for storage and analysis of biomedical data.
- Computational biology deals with mathematical modelling and simulations for understanding biological systems.
- Computational thinking is more important than computer programming in developing bioinformatics and computational biology skills.

Suggested Reading

Moody G (2004) Digital code of life: how bioinformatics is revolutionizing science, medicine, and business. Wiley, London

Mount D (2004) Bioinformatics: sequence and genome analysis, 2nd edn. Cold Spring Harbor Laboratory Press, Cold Spring Harbor

Gauthier J, Vincent AT, Charette SJ, Derome N (November 2019) A brief history of bioinformatics. Briefings in Bioinformatics 20(6):1981–1996

Biological Databases

2

Learning Objectives
You will be able to understand the following after reading this chapter:

- Different types of biological databases based on the nature of stored data.
- Disease databases focusing on an array of human genetic diseases.
- Organism-specific databases with special reference to economically important plants and animals.
- Database searching using different variants of BLAST.

2.1 Introduction

Biological data have accumulated at a faster pace in the recent past due to advent of high-throughput and cheaper next-generation sequencing technologies. Consequently, new databases are being developed in order to manage this ever increasing biological data. Databases are digital repositories based on a computerized software for storage of information in a system and their retrieval from the system using search tools. Biological databases are an important component of bioinformatics research and are usually well-annotated and cross-referenced to other databases. The primary objective of a database is to organize the data in a structured and searchable form allowing easy retrieval of useful data. The simplest form of a database is a single file containing multiple entries of same set of information. Currently, there are more than thousand biological databases providing access to multifarious omics data to biologists. Biological databases are classified based on different criteria such as levels of data coverage and data curation. Based on the extent of data coverage, biological databases consist of two main categories: comprehensive and specialized databases. Comprehensive database such as GenBank includes a variety of data

B. K. Tiwary, *Bioinformatics and Computational Biology*,
https://doi.org/10.1007/978-981-16-4241-8_2

Table 2.1 Entrez databases

Database	Area	Description
PubMed	Literature	Biomedical abstracts and citations
PubMed central	Literature	Full text articles from journals
Nucleotide	Genomes	DNA and RNA sequences
Genome	Genomes	Genome sequencing projects of different species
Gene	Genes	Detailed information on gene loci
GEO profiles	Genes	Gene expression profiles
HomoloGene	Genes	Homologous genes from different species
Protein	Proteins	Protein sequences
Structure	Proteins	Biomolecular structures
PubChem compound	Chemicals	Chemical information of compounds with structures
Online Mendelian inheritance in man (OMIM)	Genes	A catalogue of human genes and genetic disorders including phenotypes and linkage data
BioProject	Diverse data	Comprehensive collection of research studies including diverse data types
BioSample	Diverse data	A resource of annotated biological samples from diverse studies
LitCovid	Literature	A COVID-19-specific curated literature database

collected from numerous species. On the other hand, specialized databases contain data from one particular species, for example, WormBase contains data on nematode worm. Similarly, biological databases are classified into two groups, namely primary databases and secondary databases, based on the levels of data curation. Primary databases such as GenBank are created from experimentally derived raw data generated and submitted by experimental biologists. On the other hand, secondary databases such as Ensembl (maintained at the EMBL-EBI, UK), UCSC Genome Browser (maintained at the University of California, Santa Cruz, USA) and TIGR (maintained at the Institute of Genomic Research, Maryland, USA) are highly curated and are usually created from analysis of various sources of primary data. Some databases such as UniProt have characteristics of both primary and secondary databases. For example, Uniprot database stores peptide sequences generated from sequencing experiments and computationally inferred from genomic data as well.

Since majority of biological databases do not have a complete information, an integrated database retrieval system like Entrez maintained by NCBI provides integrated access to 35 distinct databases. These databases may be grouped into six areas: Literature, Genomes, Genes, Proteins, Health and Chemicals (Table 2.1). It can be searched using Boolean phrases (AND, OR and NOT) and useful data is downloaded in various formats. For example, most common Entrez databases include PubMed, SRA, Nucleotide, Protein and Structure. Similarly, Sequence Retrieval System (SRS) also provide integrated database retrieval system regarding a search term. BioStudies (www.ebi.ac.uk/biostudies) is a recent public multimodal database launched by EMBL-EBI in 2017 and it currently holds metadata descriptions from various biological studies with a plan to archive all functional genomics data in near future.

2.2 Types of Biological Databases

2.2.1 Sequence and Structure Databases

(a) *Nucleic acid databases*: All published DNA and RNA sequences are usually deposited in three parallel public databases, namely GenBank (Fig. 2.1) (available at the National Centre for Biotechnology Information), EMBL (European Molecular Biology Laboratory) and DDBJ (DNA Data Bank of Japan available at the National Institute of Genetics), maintained in the USA, the UK and Japan, respectively. These three public databases exchange their data under the International Nucleotide Sequence Database Collaboration (INSDC) framework. The important nucleotide databases are listed in Table 2.2. GenBank has grown exponentially since its inception in 1982 and currently contains more than 2.1 billion nucleotide sequences (Fig. 2.1). The current exponential increase with a doubling time of 20.8 months is very close to the doubling time of 18 months as proposed by Moore's law. There was an exponential increase in the number of taxa as well in the publically available databases (Fig. 2.2). The European Nucleotide Archive (ENA) is maintained by EMBL and provides open access to a wide range of nucleotide sequences from raw reads to finished genome sequences. The DDBJ Sequence Read Archive (DRA), maintained by the DDBJ, provides free access to raw read data and assembled genomic data from next-generation sequencing platforms. The DNA and RNA sequences are directly submitted by the researchers in these databases and are

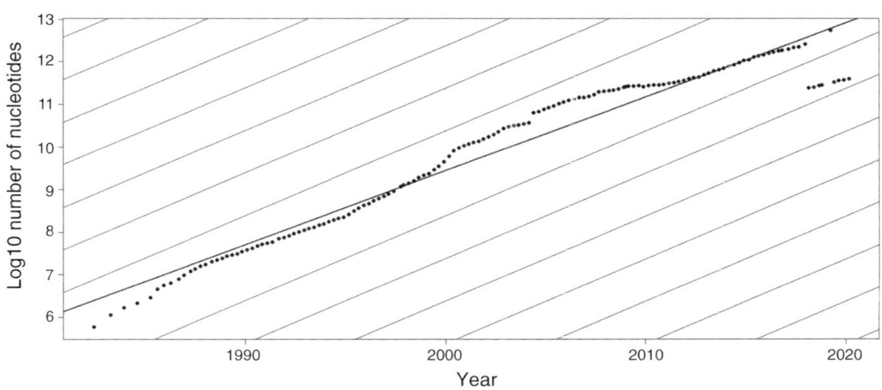

Fig. 2.1 The growth of nucleotide sequences on a logarithmic scale in public databases (GenBank/ENA/DRA) from June 1982 to March 2020. The central line is the best least square fit line representing doubling time of 20.8 months, whereas external lines indicate a doubling time of 18 months

Table 2.2 Nucleotide databases

Database	Description	URL
NCBI GenBank	Genetic sequence database	www.ncbi.nlm.nih.gov/genbank
Ensembl	A genome browser of vertebrates	www.ensembl.org
NCBI Refseq	A collection of non-redundant and well-annotated genomic sequences, transcripts and proteins	www.ncbi.nlm.nih.gov/refseq
UCSC genome browser	Interactive genome visualization browser	https://genome.ucsc.edu/
1000 genomes	A catalogue of human genomic variation	www.internationalgenome.org
GeneCards	An integrative database of predicted and annotated human genes	www.genecards.org
lncRNAdb	Database containing annotated long non-coding RNAs in eukaryotes	rnacentral.org/expert-database/lncrnadb
miRBase	Database of published miRNA sequences along with annotation	www.mirbase.org
DIANA-TarBase	A database of experimentally supported miRNA targets.	www.microrna.gr/tarbase

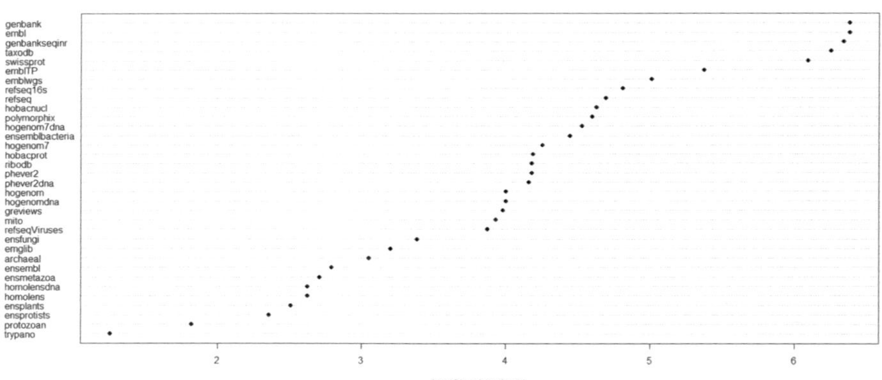

Fig. 2.2 The exponential growth in the number of taxa in the available public databases

cross-submitted in other two databases as well if submitted in one database. Each sequence in the database has a unique accession number and a version number which is common for all three databases. The GenBank (Fig. 2.3) is searched online using Entrez, whereas both EMBL and DDBJ databases are searched through SRS servers. Expressed Sequence Tags (ESTs), small fragments of mRNA with high error rate are also available on nucleic acid databases. Similarly, Genome Survey Sequences (GSS), a single-pass fragments of genomic sequences with high error rate are also integral part of these

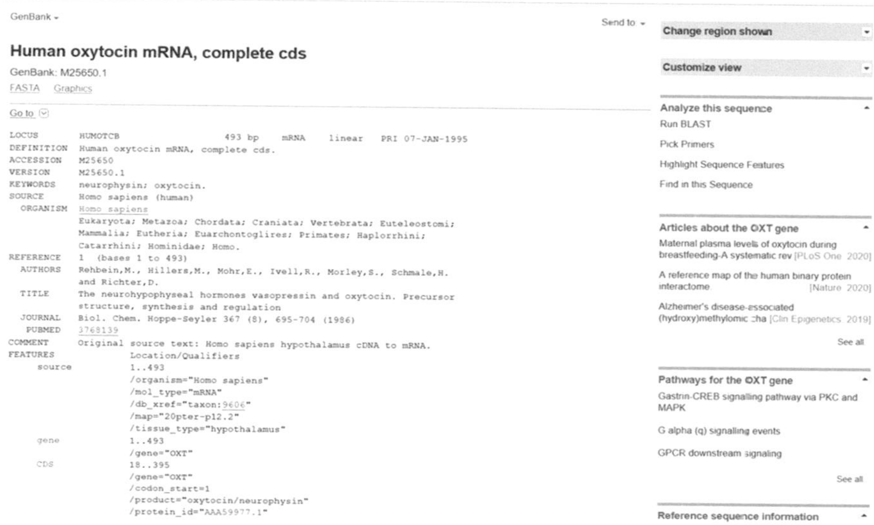

Fig. 2.3 Partial view of GenBank home page showing information on human oxytocin mRNA

databases. Moreover, whole-genome sequences of various species are also deposited in the nucleic acid databases and are regularly updated with the release of new genomic sequences. Nucleic acid databases are primarily customized for human DNA and RNA sequences useful in biomedical research. For example, a reference human genome in form of NCBI Refseq database is created and human genetic variation is also profiled in a database known as dbSNP. Ensembl is an integrated platform for genome annotation and distribution of genomic data with comprehensive annotation of genomic variants, transcript structures and regulatory regions (Fig. 2.4). It provides a valuable resource for evolutionary studies using large-scale comparative genomics data of 227 vertebrate and model species. In addition, it also provides a genome browser showing genomic variations emerging frequently in SARS-CoV2 virus (https://covid-19.ensembl.org). The University of California Santa Cruz (UCSC) Genome Browser currently provides a web-based view of 211 genome assemblies of more than hundred species (Fig. 2.5). The SARS-CoV-2 Genome Browser is recently added to UCSC Genome Browser including datasets from major annotation databases. There are many databases dedicated for various non-coding RNAs such as micro RNA (miRNA) and long non-coding RNA (lnc RNA) and a popular database for unified access to non-coding RNA sequences is RNAcentral. DIANA-TarBase is a reference database of experimentally tested numerous miRNA targets providing cell-specific miRNA-gene interaction.

(b) *Protein databases*: The protein sequences available in the protein databases (Table 2.3) are obtained from protein sequencing methods such as Edman degradation and peptide mass spectrometry. In addition, they are also inferred from three-dimensional structures obtained through X-ray crystallography and

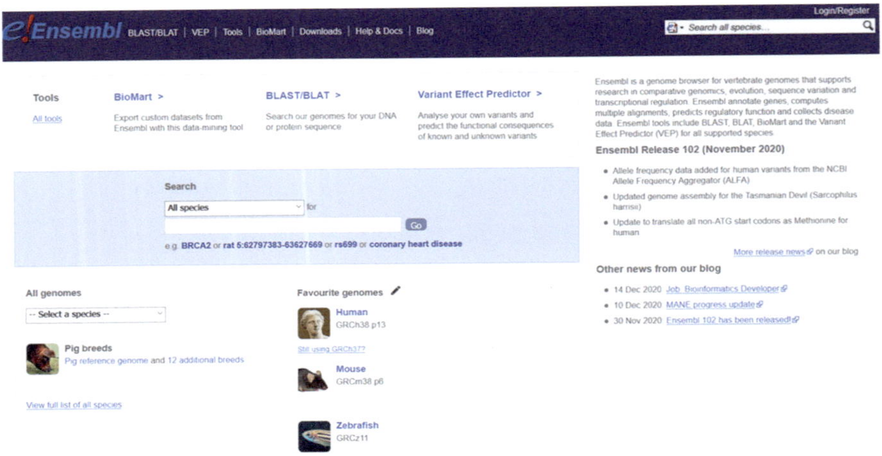

Fig. 2.4 Partial view of Ensembl genome database

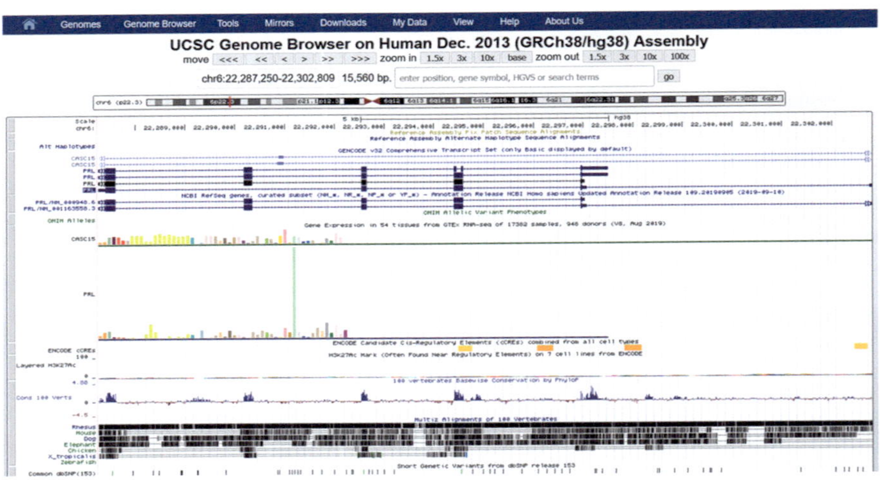

Fig. 2.5 Partial view of UCSC Genome browser

NMR. Moreover, a significant amount of protein sequence data is also obtained from translating DNA and RNA sequences. Universal protein resource (UniProt) provides a comprehensive access to high-quality protein sequences (Fig. 2.6). UniProt knowledge base (UniProtKB) is the primary source of universal protein sequence information containing more than 189 million sequences obtained from experimental sequencing as well as translated ORF sequences from EMBL. This database is endowed with well-annotated protein

Table 2.3 Protein databases

Database	Description	URL
UniProt	A public database of protein sequences with their functional information	www.uniprot.org
CATH-Gene3D	Database of protein classification and prediction of domain structure	www.cathdb.info
GenPept	Translated coding sequences from GenBank	www.ncbi.nlm.nih.gov/protein
DIP	A database of experimentally determined protein–protein interactions	http://dip.doe-mbi.ucla.edu
HPRD	Integrated platform depicting various information regarding human proteome	www.hprd.org
InterPro	Integrated database providing functional analysis of proteins	www.ebi.ac.uk/interpro
ModBase	Database containing theoretically calculated protein structure models	modbase.compbio.ucsf.edu
RCSB protein data Bank (PDB)	Public database containing experimentally determined protein and other macromolecule structures	www.rcsb.org
Pfam	Database of large collection of protein families	pfam.xfam.org
PROSITE	Database of protein domains, families and functional sites	prosite.expasy.org
ProteomicsDB	A multi-omics database including proteomics, transcriptomics and cell line viability data	www.ProteomicsDB.org
CoV3D	A database of coronavirus protein structures	cov3d.ibbr.umd.edu
STRING	A database of protein–protein interactions	string-db.org

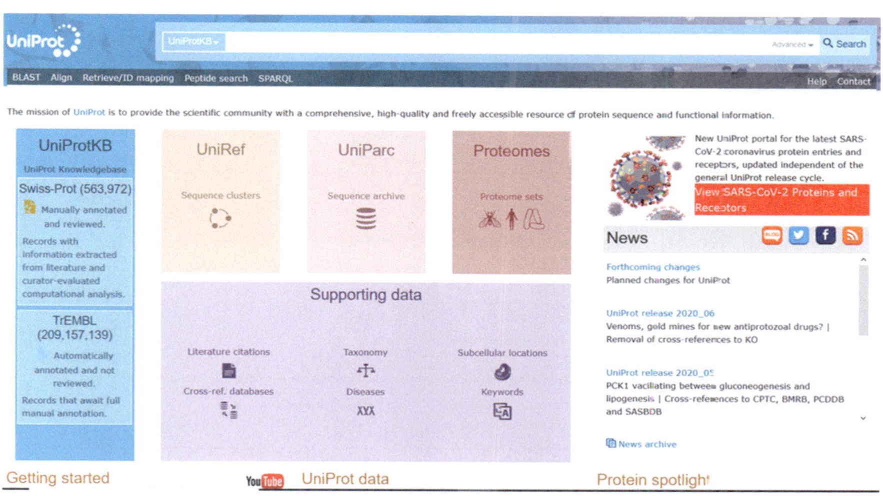

Fig. 2.6 Partial view of UniProt home page

sequences and a preliminary assignment of motifs present in the sequences. It is also cross-referenced with other useful databases. Uniprot database consists of two divisions: SwissProt and TrEMBL. TrEMBL is an automated database requiring minimum human intervention, whereas Swissprot is a well-curated database having manual entry of useful information from available literature. Other important protein databases are CATH-Gene3D, GenPept, PRF and DAD. The CATH-Gene3D database has the classification of about 151 million protein domains into 54,881 superfamilies and prediction of structural domains on the publicly available protein sequences. GeneBank Gene Products Databank (GenPept) and DAD contain translated coding sequences from GenBank and DDBJ, respectively. The Database of Interacting proteins (DIP) has a detailed documentation of experimentally determined protein–protein interactions involved in a biological process. The Human Protein Reference Database (HPRD) is an integrated platform regarding the domain architecture and post-translational modifications of proteins available in the human proteome along with their interactions and association with diseases. Pfam is a protein database dedicated to protein families and domains. The Protein Data Bank (PDB) is an international consortium consisting of four partners, namely Research Collaboratory for Structural Bioinformatics Protein Data Bank (RCSB-PDB), Protein Data Bank in Europe (PDBe), Protein Data Bank Japan (PDBj) and Biological Magnetic Resonance Bank (BMRB) (Fig. 2.7). It is a global archive for protein structures and other macromolecules determined using X-ray crystallography and NMR. This global database was launched in 1971 and contains more than 150,000 macromolecular structures adding about 12,000 new structures each year. ProteomicsDB is a protein-centric database having large quantities of human proteomics data generated using mass spectrometry. It

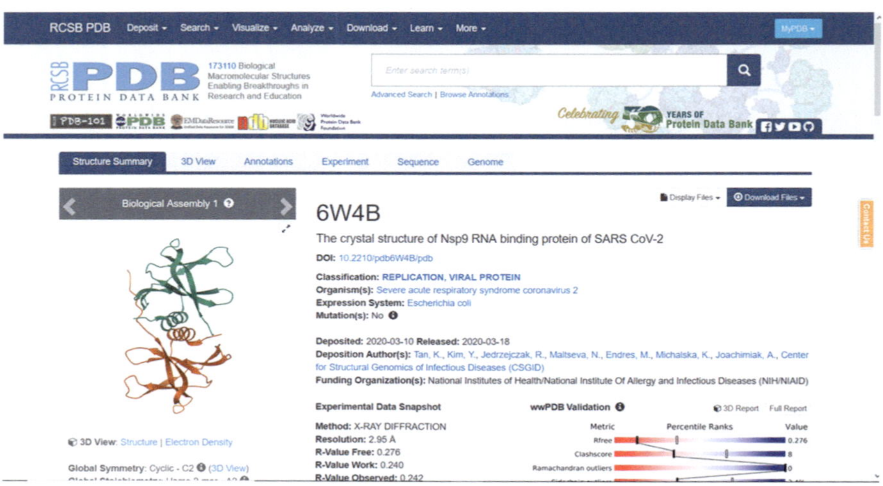

Fig. 2.7 A screenshot of RCSB-PDB home page

provides a real-time exploration of protein abundance in different tissues, body fluids and cell lines. In addition, a visual representation of diverse drug-target interaction data is also allowed in this database. STRING database provides an integrated access to all known and predicted protein–protein interactions, both physical interactions and functional associations, in more than 14,000 organisms.

2.2.2 Expression Databases

There are some useful expression databases hosting gene expression data for functional genomics studies (Table 2.4). High-throughput gene expression data derived from both microarray and RNA-Seq platforms are usually archived in two public databases: NCBI Gene Expression Omnibus (GEO) and EBI ArrayExpress (AE). Gene expression Omnibus database was started in 2000 and is maintained by NCBI. The overall structure of this database consists of platform record, sample record and series record. Each GEO DataSet is an upper-level object having unique GEO accession number starting with prefix GDS. DataSets are represented both in terms of experiment-centric and gene-centric information. The gene-centric information provides quantitative expression of a single gene across DataSets and is known as *GEO Profile*. In addition to gene expression data, GEO contains other functional genomic data such as copy number variations and transcriptional factor binding. Another public database ArrayExpress is maintained by EBI and was launched in

Table 2.4 Expression databases

Database	Description	URL
NCBI GEO	Gene expression data from microarray and sequencing.	www.ncbi.nlm.nih.gov/geo
EBI ArrayExpress	Gene expression data from microarray and sequencing.	www.ebi.ac.uk/arrayexpress
Expression atlas	Gene and protein expression data from various species and conditions.	www.ebi.ac.uk/gxa
Gene expression archive	Gene expression data from microarray and sequencing.	www.ddbj.nig.ac.jp/gea
Human protein atlas	A map of all human proteins in cells, tissues and organs	www.proteinatlas.org
ONCOMINE	Gene expression and sample data from different cancer types.	www.oncomine.org
TiGER	Tissue-specific gene expression profile and gene regulation data	bioinfo.wilmer.jhu.edu/tiger
DualSeqDB	A host-bacterial pathogen RNA-sequencing database having combined gene expression data during infection process.	www.tartaglialab.com/dualseq
LncExpDB	An expression database of human long non-coding RNAs	https://bigd.big.ac.cn/lncexpdb

2002. It consists of three components: ArrayExpress Repository, ArrayExpress Warehouse and ArrayExpress Atlas. The DDBJ recently started a similar database known as Genomic Expression Archive (GEA) for hosting the functional genomics data. All three of the public databases follow the guidelines of reporting standards the Minimum Information About a Microarray Experiment (MIAME) for microarray data and the Minimum Information about a high-throughput Nucleotide SEQuencing Experiment (MINSEQE) for sequencing data. Expression atlas of EBI provides information regarding the gene and protein expression in different species under various biological conditions. Moreover, Human Protein Atlas (HPA) has a detailed map of all proteins expressed and located in various human cells, tissues and organs. It consists of tissue atlas, subcellular atlas, pathology atlas, blood atlas, metabolic atlas and brain atlas. Tissue-specific Gene Expression and Regulation (TiGER) provides large-scale data sets not only for tissue-specific genes but also for tissue-specific transcriptional regulatory elements. A user can retrieve useful information regarding a gene, transcriptional factors or a specific tissue from TiGER using gene view, TF view and tissue view, respectively. Oncomine is a database having heterogeneous cancer profiles and provides access to gene expression signatures and sample data of more than 500 cancer types and cancer cell lines. DualSeqDB is a resource for differential gene expression in pathogenic bacteria and their natural hosts upon infection at different time points. LncExpDB is an expression database and a comprehensive resource for expression landscape of long non-coding RNAs under various biological conditions.

2.2.3 Pathway Databases

Pathway databases provide a roadmap connecting various omics data such as genomics, transcriptomics, proteomics, metagenomics and metabolic data (Table 2.5). Metabolic pathways database hosts curated data regarding metabolic

Table 2.5 Metabolic pathway databases

Database	Description	URL
KEGG	Database of biological systems including drugs and diseases	www.genome.jp/kegg
MetaCyc	Reference database of metabolic pathways	metacyc.org
PathBank	Metabolic and signalling pathway database of model organisms	www.pathbank.org
Reactome	Curated pathway database	reactome.org
HMDB	Human metabolome database	hmdb.ca
BiGG models knowledge base	A database of genome-scale metabolic models	bigg.ucsd.edu
Plant Reactome	Comparative plant pathway database	plantreactome.gramene.org
Ingenuity pathway analysis (IPA)	Commercial pathway database developed by Qiagen	digitalinsights.qiagen.com

pathways from all domains of life along with chemical compounds, reactions and enzymes. They are useful in complex bioinformatics analysis such as flux balance analysis and cellular modelling. KEGG and MetaCyc are the oldest and most popular metabolic databases. MetaCyc is the largest reference database of experimentally determined metabolic pathways and enzymes from all domains of life. It contains 2749 pathways currently extracted from thousands of publications. PathoLogic, a component of Pathway Tools program, performs computational prediction of metabolic pathways in an organism using this reference database. The major limitations of this database are lack of any information regarding protein signalling, disease and developmental process pathways. KEGG (Kyoto Encyclopaedia of Genes and Genomes) is a comprehensive functional knowledge base of genes and genomes at the molecular level and higher biological levels as well. KEGG Orthology (KO) database hosts the molecular functions of various genes and proteins. The higher-level biological functions are represented as KEGG pathway maps describing the biological networks of molecular interactions and metabolic reactions. In addition, the KEGG database also provides chemical information regarding the metabolites and other small molecules in form of KEGG LIGAND and health-related information of drugs and diseases in form of KEGG MEDICUS. However, KEGG database does not provide information regarding lipid synthesis, cellular location and metabolite signalling. PathBank is a comprehensive pathway database of ten model organisms including human, rat, cow, yeast, etc. It provides a detailed account of map of metabolic pathways and signalling, protein signalling, disease, drug and physiological pathways in these model organisms. Reactome is a relational database of signalling, transcriptional regulation, disease and metabolic pathways. It provides a bioinformatics tool for analysis of pathways for systems biology research. The human metabolome database (HMDB) is a comprehensive electronic knowledge base of small-molecule metabolites present in human body along with hyperlinks to other useful databases such as KEGG, MetaCyc and PDB. BiGG models knowledge base is a high-quality genome-scale metabolic model repository containing more than 100 BiGG models of different organisms including a human-specific model known as Recon3D. In addition, there are 515 strain-specific draft genome-scale models across three organisms. Plant Reactome is a comparative and systems-level plant pathway database using rice as a reference species and the pathway knowledge derived from rice is extended to other 82 plant species using gene-orthology prediction. A total of 298 reference pathways are hosted by this database related to metabolic pathways, transcriptional networks, hormone signalling pathways and developmental processes. Ingenuity Pathway Analysis (IPA) is a commercial database developed by QIAGEN for visualization and analysis of omics data using a network or pathway.

2.2.4 Disease Databases

Disease database is one of the most important resource for biomedical research (Table 2.6). The Online Mendelian Inheritance in Man (OMIM) of NCBI contains

Table 2.6 Disease databases

Database	Description	URL
Online Mendelian inheritance in man (OMIM)	Comprehensive compendium of human genes and genetic phenotypes	www.omim.org
The Cancer genome atlas (TCGA)	A cancer database having genomic, transcriptomic, proteomic and epigenomic data	www.cbioportal.org
Human gene mutation database (HGMD)	Database of all published gene mutations involved in human inherited diseases	www.hgmd.cf.ac.uk
GWAS central	Comprehensive repository of genome-wide association study data	www.gwascentral.org
HbVar	Database of human haemoglobin variants and thalassemia mutations	globin.cse.psu.edu/globin/hbvar/
MalaCards	Integrated database of human diseases	www.malacards.org
miR2Disease	A comprehensive database on miRNA-disease relationship	www.mir2disease.org
DisGeNET	Database of genes and variants associated with human diseases	www.disgenet.org
STAB	A cell atlas providing cellular landscape of human brain and neuropsychiatric diseases.	http://stab.comp-sysbio.org
canSAR	A knowledge base for cancer translational research and drug discovery	http://cansar.icr.ac.uk
CNCDatabase	Cornell non-coding Cancer driver database is a manually curated database regarding non-coding cancer drivers	https://cncdatabase.med.cornell.edu/
PAGER-CoV	Pathways, annotated gene-lists and gene signatures electronic repository for Corona virus	discovery.informatics.uab.edu/PAGER-CoV/

useful information regarding human genes and genetic diseases along with gene sequences and associated phenotypes. It provides a complete description regarding a disease gene, the phenotypes associated with the disease gene and other genes associated with the disease. The Cancer Genome Atlas (TCGA) is joint programme of National Cancer Institute and National Human Genome Research Institute launched in 2006. It hosts more than 2.5 pentabytes of genomic, epigenomic, transcriptomic and proteomic data of 33 cancer types. canSAR is a largest public resource of cancer multidisciplinary data for integrative translational research and drug discovery. Similarly, CNCDatabase provides a detailed information about predicted non-coding cancer drivers located on promoters, untranslated regions, enhancers and non-coding RNAs. The Human Gene Mutation Database (HGMD) is a collection of all published gene mutations involved in human inherited diseases. The GWAS Central database is most comprehensive repository of genome-wide association study (GWAS) data related to more than 1400 phenotypes. It provides integrative access and graphic visualization of GWAS data collected from more than 3800 studies. The HbVar is a database of human genomic sequence changes not only involved in the generation of different variants of haemoglobin but also causing different forms of haemoglobinopathies and thalassemia. A detailed information is provided regarding genomic sequence alterations along with biochemical, haematological and pathological changes. The MalaCards human disease database is an integrated database of human diseases having a collection of more than 21,000 diseases. The miR2Disease is a manually curated database hosting detailed information on a microRNA-disease relationship. The current version of the database accounts for 3273 relationship between 349 miRNA and 163 diseases. The DisGeNET is a knowledge platform providing data regarding disease-associated genes and their variants collected from various sources including the scientific publication. The current version of this comprehensive database hosts more than 17,000 genes and 117,000 genomic variants. STAB is a spatio-temporal cell atlas of regional cellular heterogeneity of human brain providing a landscape of transcriptome dynamics during neuropsychiatric disorders. PAGER-CoV is a newly developed web-based disease database to interpret the functional genomics studies of coronavirus disease such as inflammatory response and organ damage in the host.

2.2.5 Organism-Specific and Virus Databases

There are specialized organism-specific databases available for some model animals, economically important plants and animals (Table 2.7). In addition, specific genomic databases of pathogenic viruses are also available. The Alliance of Genome Resources Portal (www.alliancegenome.org) provides an integrated access to genome of diverse model species used to study human biology. WormBase which is one of the founding member of this alliance is a repository of detailed biological

Table 2.7 Organism-specific and Virus databases

Database	Description	URL
WormBase	Database of experimental data of nematode *C. elegans*	wormbase.org
SilkDB	Database of silkworm genome	silkdb.bioinfotoolkits.net
MBKbase-rice	Integrated omics database of rice	www.mbkbase.org/rice
Bovine genome database (BGD)	Bovine genomics database	bovinegenome.org
Pig genome database (PGD)	Pig genomics database	www.animalgenome.org/pig/genome/db
Zebrafish information network (ZFIN)	Zebrafish genomics database	zfin.org
GISAID	A global database of influenza viruses and SARS-CoV-2 virus	www gisaid.org
GESS	A database of global evaluation of SARS-CoV-2/hCoV-19 sequences	wan-bioinfo.shinyapps.io/GESS

information from experimental observations from more than 1400 laboratories across the world on multiple nematode species. Sericulture is an important component of agriculture and silkworm is an insect species which laid the foundation of sericulture. SilkDB is a comprehensive database of silkworm genome containing high-quality chromosome level assembly along with various visualization tools. MBKbase is an integrated molecular breeding knowledge base of important crop genome sub-databases for rice, soybean, wheat and maize. The rice sub-database of this knowledge base provides rice germplasm information including pan-genomic view of multiple reference genomes. There are genomic databases for economically important livestock species useful for research and molecular breeding of these species. The Bovine Genome Database (BGD) provides genome browsing, genome annotation and data mining of the bovine genome. In addition, this database also offers bovine sequence searching using BLAST. The Pig Genome Database (PGD) is a rich collection of quantitative trait loci (QTL), whole-genome association study (WGAS), candidate genes and gene expression in the pig genome. Zebrafish is a popular model organism for developmental biology and toxicological research. The Zebrafish Information Network (ZFIN) has a wide variety of experimental data related to the zebrafish research including genetic and genomic data. The genome analysis is essential to understand the evolutionary history of pathogenic bacterial strains and different variants of pathogenic viruses. The genomic database of these pathogens is a useful resource for understanding the outbreak of any epidemic or pandemic disease. For example, the recent pandemic caused by human CoV-19 can be better understood and prevented using its genomic diversity across the world. In this connection, the GISAID (Global Initiative on Sharing All Influenza data) database provides open access to the genomic sequences of all influenza viruses and human SARS-CoV-2. Moreover, Global Evaluation of SARS-CoV-2/hCoV-19 Sequences (GESS) is a web tool providing analysis of single-nucleotide variants (SNV) collected from more than 300,000 high-quality and high-coverage human CoV-19 viral genomes from GISAID database. It allows a user to search and download SNVs at any single or multiple viral genomic position(s) from any particular country or region and is a useful resource to monitor the migration and evolution of CoV-19 virus. Moreover, CoV3D is a comprehensive resource for coronavirus protein structures such as spike glycoprotein and their complexes with antibodies, receptors and small molecules and provides useful information for structure-based design of drugs and vaccines.

2.3 Database Searching

The Basic Local Alignment Search Tool (BLAST) is most widely used search tool for sequence databases. It finds a region of local similarity (conserved sequence pattern) between a query DNA or protein sequence against a target database. The ultimate aim of a BLAST program is to infer an evolutionary and functional relationship between two DNA or protein sequences. The original version of BLAST was launched in 1990 followed by development of various variants of

Table 2.8 Different variants of NCBI-BLAST program

BLAST program	Description
BLASTN	A nucleotide sequence is searched against a nucleotide sequence database
BLASTP	A protein sequence is searched against a protein sequence database
BLASTX	All reading frames of a nucleotide sequence is against a protein sequence database
TBLASTN	A protein sequence is searched against all reading frames of a nucleotide sequence database
TBLASTX	Six-frame translations of nucleotide sequence is searched against six-frame translations of a nucleotide sequence database
MEGABLAST	Suitable for finding alignment between closely related sequences
PSI-BLAST	Suitable for searching remote homologs or members of a protein family
PHI-BLAST	Suitable for finding protein sequences with specific pattern in the database

BLAST specialized for different types of databases. Table 2.8 describes various variants of BLAST along with their applications. There are three important aspects of a search process: the input query sequence, target database and choosing a customized BLAST program. For example, PSI-BLAST is a useful program to identify remote homologs in different species. BLAST finds a local alignment in each match in the database above a certain threshold S. These matches are known as high scoring segment pairs (HSPs) along with their E-values. An E-value of 10^{-5} is usually taken as cut-off during a BLAST search underlying the fact that the score or a greater value than the score is likely to occur by chance in one out of 100,000 matches. The Position-Specific Iterated (PSI)-BLAST searches remote protein homologs in a protein database in multiple iterative steps using position-specific scoring matrix. FASTA is a similarity search program like BLAST and its input sequence format is widely known as FASTA format. Blast-like alignment tool (BLAT) is another efficient program to find sequences of greater similarity (more than 95%) but has less sensitivity to divergent sequences. The BLAT algorithm keeps the index of complete genome in the computer memory requiring 2GB RAM. However, sensitivity of this program is significantly inferior to NCBI-BLAST.

2.4 Exercises

1. Aromatase, a member of Cytochrome P450 family of enzymes, performs the conversion of androgens to oestrogens in brain, testes and ovary across different species of vertebrates. Teleost fishes have two structurally and functionally different forms of aromatase, aromatase A and aromatase B. Find the remote homologs of human aromatase in different species of teleost fishes using an appropriate variant of BLAST program.

Solution

1. Retrieve the protein sequence of the human aromatase from GenBank or Ensembl.
2. Use this protein sequence as a query sequence to perform PSI-BLAST (Figs. 2.8, 2.9, and 2.10).
2. The *TP53* gene encodes for a tumour suppressor protein responding to various kind of cellular stresses. It plays an important role in the prevention of cancer and is also known to be mutated in majority of human cancer. Find the location and number of alternative transcripts of *TP53* gene in human and infer the

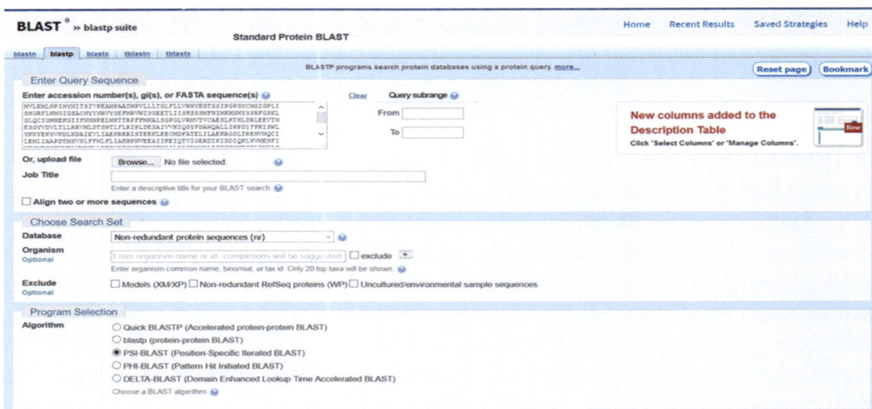

Fig. 2.8 PSI-BLAST submission page using default parameters

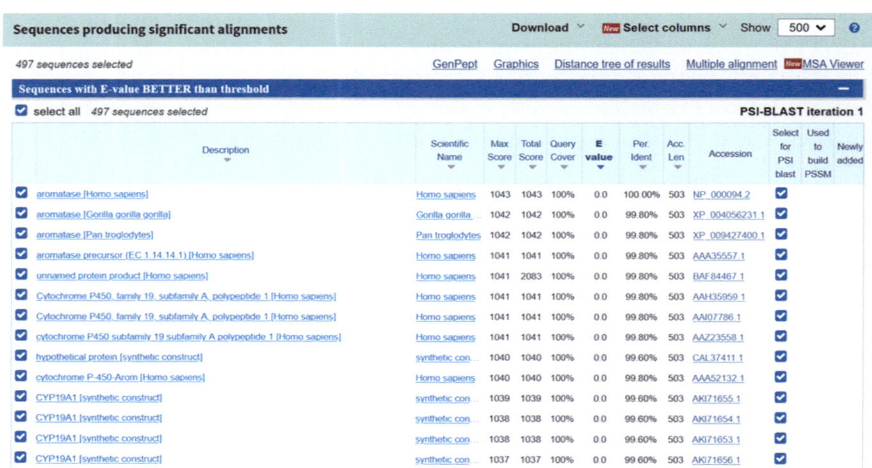

Description	Scientific Name	Max Score	Total Score	Query Cover	E value	Per. Ident	Acc. Len	Accession	Select for PSI blast	Used to build PSSM	Newly added
aromatase [Homo sapiens]	Homo sapiens	1043	1043	100%	0.0	100.00%	503	NP_000094.2	☑		
aromatase [Gorilla gorilla gorilla]	Gorilla gorilla ...	1042	1042	100%	0.0	99.80%	503	XP_004056231.1	☑		
aromatase [Pan troglodytes]	Pan troglodytes	1042	1042	100%	0.0	99.80%	503	XP_009427400.1	☑		
aromatase precursor (EC 1.14.14.1) [Homo sapiens]	Homo sapiens	1041	1041	100%	0.0	99.80%	503	AAA35557.1	☑		
unnamed protein product [Homo sapiens]	Homo sapiens	1041	2083	100%	0.0	99.80%	503	BAF84467.1	☑		
Cytochrome P450, family 19, subfamily A, polypeptide 1 [Homo sapiens]	Homo sapiens	1041	1041	100%	0.0	99.80%	503	AAH35959.1	☑		
Cytochrome P450, family 19, subfamily A, polypeptide 1 [Homo sapiens]	Homo sapiens	1041	1041	100%	0.0	99.80%	503	AAI07786.1	☑		
cytochrome P450 subfamily 19 subfamily A polypeptide 1 [Homo sapiens]	Homo sapiens	1041	1041	100%	0.0	99.80%	503	AAZ23558.1	☑		
hypothetical protein [synthetic construct]	synthetic con...	1040	1040	100%	0.0	99.60%	503	CAL37411.1	☑		
cytochrome P-450-Arom [Homo sapiens]	Homo sapiens	1040	1040	100%	0.0	99.80%	503	AAA52132.1	☑		
CYP19A1 [synthetic construct]	synthetic con...	1039	1039	100%	0.0	99.60%	503	AKI71655.1	☑		
CYP19A1 [synthetic construct]	synthetic con...	1038	1038	100%	0.0	99.60%	503	AKI71654.1	☑		
CYP19A1 [synthetic construct]	synthetic con...	1038	1038	100%	0.0	99.60%	503	AKI71653.1	☑		
CYP19A1 [synthetic construct]	synthetic con...	1037	1037	100%	0.0	99.60%	503	AKI71656.1	☑		

Fig. 2.9 Part of the Iteration 1 table for the human aromatase search showing closely related sequences

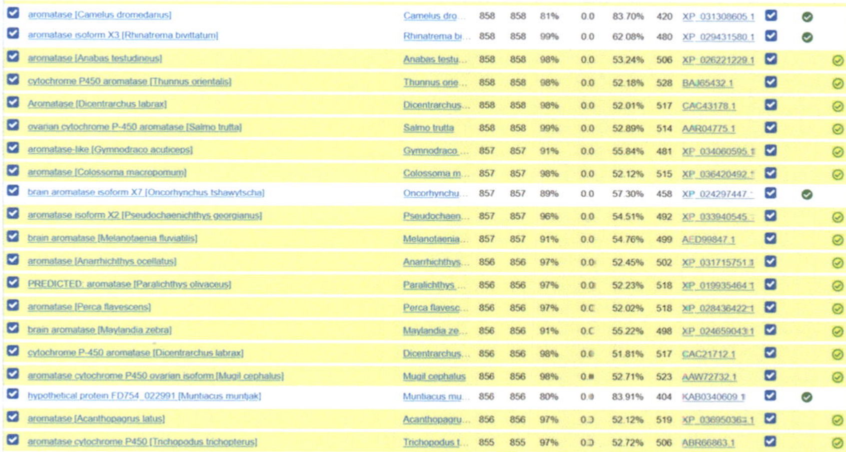

Fig. 2.10 New remote homologs are identified with each iteration of PSI-BLAST search. Part of the interaction table shows newly identified fish homologous sequences (highlighted in yellow)

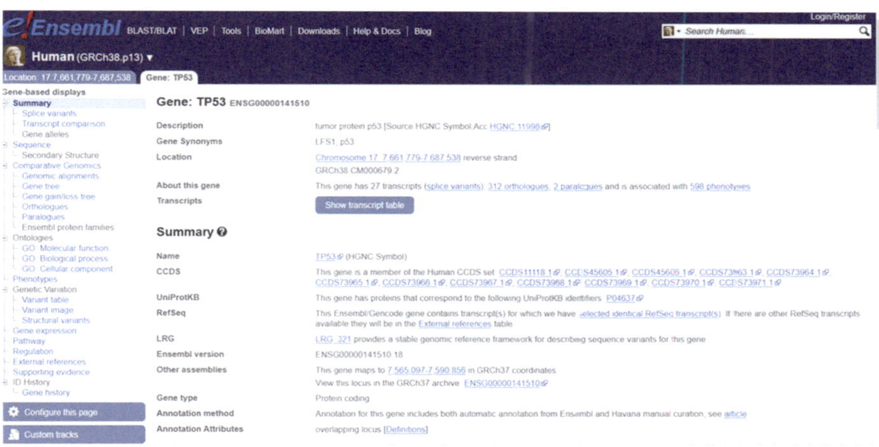

Fig. 2.11 Partial view of Ensembl genome browser showing details regarding human *TP53* gene

evolutionary history of *TP53* gene gain and loss in different genomes available in the Ensembl database.

Solution

1. Search the term "*TP53* gene" under human species in the Ensembl database.
2. It is located on the reverse strand of Chromosome 17: 7,661,779-7,687,538 (Fig. 2.11).
3. The *TP53* gene consists of 27 alternative transcripts in human (Fig. 2.12).

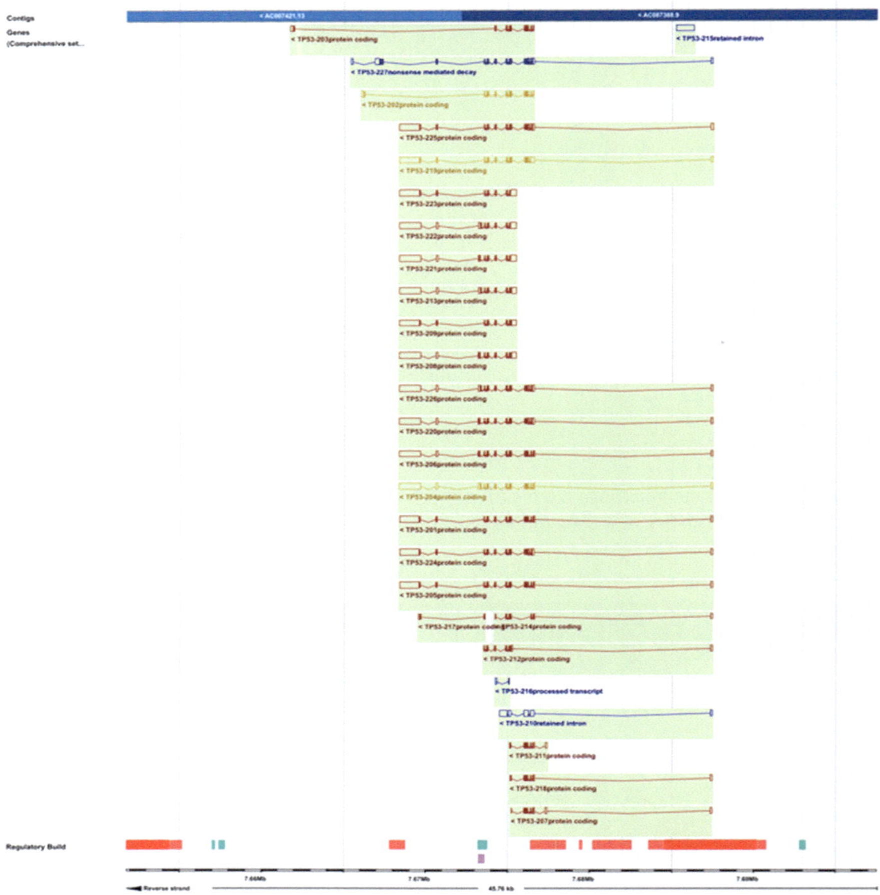

Fig. 2.12 The Ensembl genome browser showing 27 alternative transcripts of *TP53* gene

4. View the gene gain/loss tree to see the evolutionary history of human *TP53* gene (Fig. 2.13). The species tree shows significant gene gain events or expansions (indicated in red branch colour), gene loss events or contractions (indicated in green branch colour) and no changes (indicated in blue branch colour). Each node indicates the number of gene family present in the respective extant or extinct species. In *TP53* gene family tree, two species, namely Siamese fighting fish (*B. splendens*) and African elephant (*L. africana*), have undergone remarkable expansion in form of 27 members and 14 members, respectively. In addition, 16 species show significant expansions and three species show significant contractions.

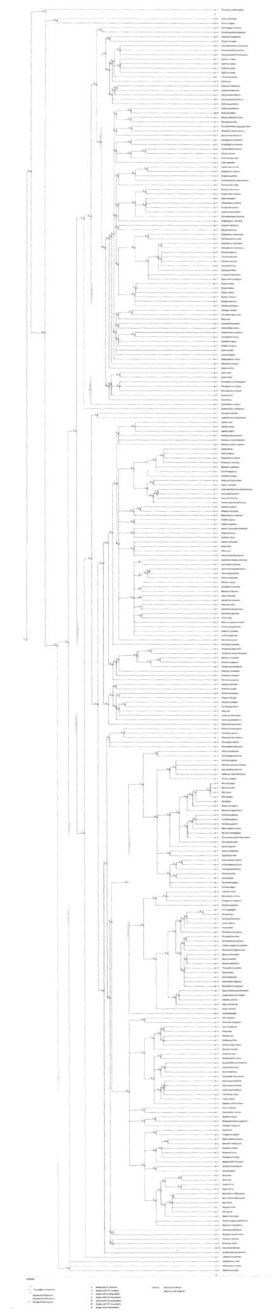

Fig. 2.13 The gene gain/loss tree of *TP53* gene

2.5 Multiple Choice Questions

1. Which of the following is a specialized database?
   (a) GenBank
   (b) EMBL
   (c) WormBase
   (d) DDBJ
2. Which of the following database is a secondary database?
   (a) Ensembl
   (b) GenBank
   (c) DDBJ
   (d) EMBL
3. The database retrieval system of NCBI is known as:
   (a) Entrez
   (b) SRS
   (c) GenBank
   (d) BLAST
4. A unified access to non-coding RNA data is provided by:
   (a) RNAcentral
   (b) Genome
   (c) Gene
   (d) GEO profiles
5. The primary source of universal protein sequences is:
   (a) UniProtKB
   (b) ProteomicsDB
   (c) HPRD
   (d) PROSITE
6. Which of the following is NOT a partner in the Protein Data Bank consortium?
   (a) RCSB
   (b) PDBe
   (c) BMRM
   (d) HPRD
7. A repository for high-quality genome-scale metabolic model is known as:
   (a) BiGG models knowledge base
   (b) GSMM models knowledge base
   (c) RGMM models knowledge base
   (d) HGSM models knowledge base
8. Which of the following database is a rich resource for cancer data?
   (a) TCGA
   (b) ATGA
   (c) TGAA
   (d) ATGC

9. Which of the following database is a useful resource for human SARS-CoV-2 viruses?
   (a) GISAID
   (b) SARSV
   (c) HSCOV
   (d) DSCOV
10. Name the variant of BLAST useful for searching remote homologs of a protein:
   (a) PHI-BLAST
   (b) PSI-BLAST
   (c) MEGABLAST
   (d) TBLASTX

Answer: 1. c 2. a 3. a 4. a 5. a 6. d 7. a 8. a 9. a 10. b

Summary

- Biological database provides a structured platform for retrieval of useful data.
- Entrez is an integrated database retrieval system for 35 diverse databases.
- GenBank, ENA, DDBJ and Ensembl are major nucleotide databases.
- UniProt Knowledgebase (UniProtKB) provides a comprehensive resource for protein sequences.
- Protein Data Bank (PDB) is a global archive for protein structures and other macromolecules.
- GISAID database is a comprehensive genomic resource for different variants of human SARS-CoV-2 virus.

Suggested Reading

Letovsky SI (1999) Bioinformatics: databases and systems. Springer, Boston

Lesk AM (2004) Database annotation in molecular biology: principles and practice. Wiley, Chichester

Revesz P (2010) Introduction to databases: from biological to Spatio-temporal. Springer, London

Annual nucleic acid research database issues, Oxford University Press: Oxford, 2015-2021

Statistical Computing Using R

3

Learning Objectives
You will be able to understand the following after reading this chapter:

- R environment, data structures and functions.
- Data input and output in the R environment.
- Statistical analysis of biological data.
- Common statistical tests using R.
- Creating graphs using R.

3.1 Introduction to R Language

Statistical computing is an interface of computer science and statistics for analysis of large data sets. R language is an object-oriented functional programming language used for statistical computing. Robert Gentleman and Ross Ihaka developed this program in early 1990s from the S language. It is more popular than any other commercial statistical software in biological research because it is not only open source but continuously being enriched and updated by the contributions from the experts across the globe. R can be operated in both interactive and batch modes and it stores its workspace in RAM. In interactive modes, commands are typed in succession one after other and consequently the result is displayed. However, batch modes do not require regular interaction with the user and usually run in an automated fashion for a longer duration. All R commands are executed in a window called the R console (Fig. 3.1). Even, the R codes are stored in a file with a suffix .R or .r and later executed with *source* command. The usual assignment operator in R is <- or ->, although = can also be used. The R programming deals with writing functions. A function consists of a number of instructions for computing a result from inputs.

B. K. Tiwary, *Bioinformatics and Computational Biology*,
https://doi.org/10.1007/978-981-16-4241-8_3

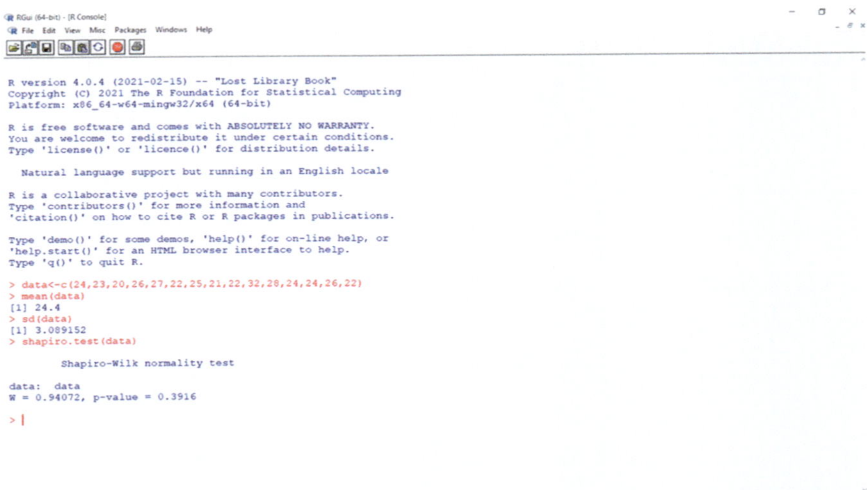

Fig. 3.1 The R console window showing implementation of some codes

3.2 Data Structures in R

The R language has a variety of data structures such as vectors, character strings, matrices, lists, data frames and classes. Vector is the R workhorse and consists of elements of similar data type such as integer or character. However, scalars or individual numbers in the R are essentially one-element vector. On the other hand, character strings are truly single-element vectors having the character mode. An R matrix consists of rectangular array of numbers. This mathematical structure has two attributes: rows and columns. An R list consists of multiple values of different data types. So, different components are packaged a single list that can be returned by an R function. The internal structure of a list can be printed by command str(), where str stands for structure. Data frame is a two-dimensional structure with rows and columns akin to a matrix. However, each column has a different mode unlike matrix. For example, one column may be numeric, whereas another column may consist of character strings in a data frame. A class defines the type of an object in connection with object-oriented programming in R. It is usually defined by the user during R programming. Generic functions operate on an object based on its class.

3.3 R Packages

R packages are organized libraries developed for different functionalities. These packages are usually downloaded from two specific repositories known as Comprehensive R Archive Network (CRAN) and Bioconductor. The standard installation

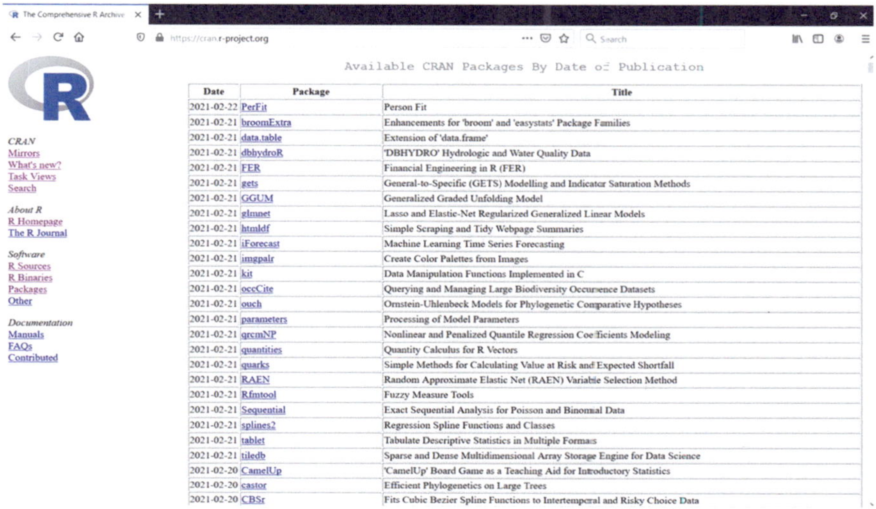

Available CRAN Packages By Date of Publication

Date	Package	Title
2021-02-22	PerFit	Person Fit
2021-02-21	broomExtra	Enhancements for 'broom' and 'easystats' Package Families
2021-02-21	data.table	Extension of 'data.frame'
2021-02-21	dbhydroR	'DBHYDRO' Hydrologic and Water Quality Data
2021-02-21	FER	Financial Engineering in R (FER)
2021-02-21	gets	General-to-Specific (GETS) Modelling and Indicator Saturation Methods
2021-02-21	GGUM	Generalized Graded Unfolding Model
2021-02-21	glmnet	Lasso and Elastic-Net Regularized Generalized Linear Models
2021-02-21	htmldf	Simple Scraping and Tidy Webpage Summaries
2021-02-21	iForecast	Machine Learning Time Series Forecasting
2021-02-21	imgpalr	Create Color Palettes from Images
2021-02-21	kit	Data Manipulation Functions Implemented in C
2021-02-21	occCite	Querying and Managing Large Biodiversity Occurrence Datasets
2021-02-21	ouch	Ornstein-Uhlenbeck Models for Phylogenetic Comparative Hypotheses
2021-02-21	parameters	Processing of Model Parameters
2021-02-21	qrcmNP	Nonlinear and Penalized Quantile Regression Coefficients Modeling
2021-02-21	quantities	Quantity Calculus for R Vectors
2021-02-21	quarks	Simple Methods for Calculating Value at Risk and Expected Shortfall
2021-02-21	RAEN	Random Approximate Elastic Net (RAEN) Variable Selection Method
2021-02-21	Rfmtool	Fuzzy Measure Tools
2021-02-21	Sequential	Exact Sequential Analysis for Poisson and Binomial Data
2021-02-21	splines2	Regression Spline Functions and Classes
2021-02-21	tablet	Tabulate Descriptive Statistics in Multiple Formats
2021-02-21	tiledb	Sparse and Dense Multidimensional Array Storage Engine for Data Science
2021-02-20	CamelUp	'CamelUp' Board Game as a Teaching Aid for Introductory Statistics
2021-02-20	castor	Efficient Phylogenetics on Large Trees
2021-02-20	CBSr	Fits Cubic Bezier Spline Functions to Intertemporal and Risky Choice Data

Fig. 3.2 CRAN, a network of ftp and webservers providing access to R codes, documentation and packages

process installs two basic packages, namely *base* and *stats* by default. CRAN provides thousands of packages for a wide range of statistical analysis and graphics (Fig. 3.2).

3.4 Data Input and Output in R

Any session in R starts with choosing its working directory. R imports a data from a datafile and builds an R object in the workspace. Here, the workspace is an abstract concept that indicates the storage of data in RAM during a specific R session. The workspace can be permanently stored in a binary file known as .RData file. In addition to .RData file, all the commands executed in a particular R session can be saved in a simple text file known as .RHistory file and reloaded again. First, the current working directory needs to be checked with the function getwd(). The current working directory may be changed through another function setwd(). R always searches any file in the working directory and writes a file in the same directory. The command line starts with a character > and continuation of a line is indicated by +. The generic function read.table reads a data from a file and writes the insides into an R object called data frame. We may add the separator in terms of tabulation characters (sep = "\t"), white spaces (sep = " ") or commas (sep = ",").

3.5 Biological Data Analysis in R

The measurement data obtained through biological experiments can be stored in form of a data vector. For example, gene expression values in five samples of a gene are stored using concatenate command c() which is as follows:

```
>gene <- c(1.25, 1.35, 1.00, 1.5, 1.75)
```

Here, an object *gene* is created including five gene expression values in different samples. The simple computations such as sum, mean, median, standard deviation and variance of these values can be computed using respective built-in functions.

```
> mean (gene)
[1] 1.37
> median(gene)
[1] 1.35
> sd(gene)
[1] 0.279732
> var(gene)
[1] 0.07825
```

However, data values are often stored in form of matrix consisting of rows and columns.

3.5.1 Statistical Distributions

In order to test a biological hypothesis, the nature of statistical distribution should be known for proper application. The important statistical distributions are binomial distribution, normal distribution, T-distribution, F-distribution and chi-squared distribution. Only the binomial distribution is a discrete distribution, whereas rest of the distributions are continuous. The binomial distribution is common in biological experiments with repeated trials with two outcomes like success/failure, healthy/disease, etc. (Figs. 3.3 and 3.4). The number of success is denoted by k and number of trials is denoted by n. The mean is np, the variance is np(1-p) and the standard deviation is square root of np(1-p) in case of a binomially distributed variable. The density function of a binomial distribution in R can be computed by

```
>dbinom(k,n,p)
```

The normal distribution of data is assumed for many statistical analyses such as gene expression values obtained using microarray data. The data values in normal distribution are represented by a bell-shaped curve or normal density with mean mu and variance sigma squared. This curve is symmetric around the mean value. We can

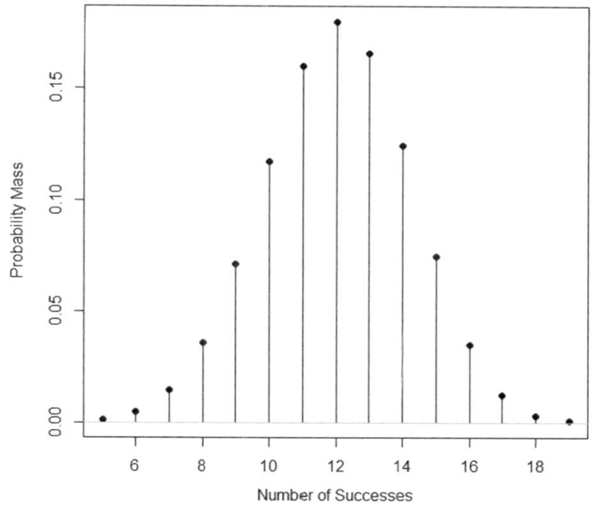

Fig. 3.3 Binomial probabilities with n = 20 and p = 0.6

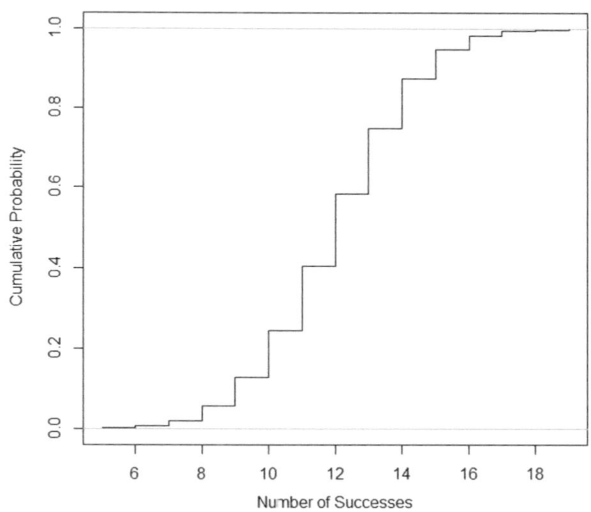

Fig. 3.4 Binomial cumulative probabilities with n = 20 and p = 0.6

draw a random sample of size 100 with the population mean of 1.5 and the population standard deviation of 0.6 using the R command

```
>rnorm(100,1.5,0.6)
```

A variable with normal distribution can be standardized into a standard normally distributed Z by subtracting mu and dividing the result with sigma. The standard normally distributed Z has a fixed mean zero and standard deviation one (Figs. 3.5 and 3.6). The chi-squared distribution Figs. 3.7 and 3.8), T-distribution (Figs. 3.9 and 3.10) and F-distribution (Figs. 3.11 and 3.12) are modifications of normal distribution and are useful in testing hypothesis.

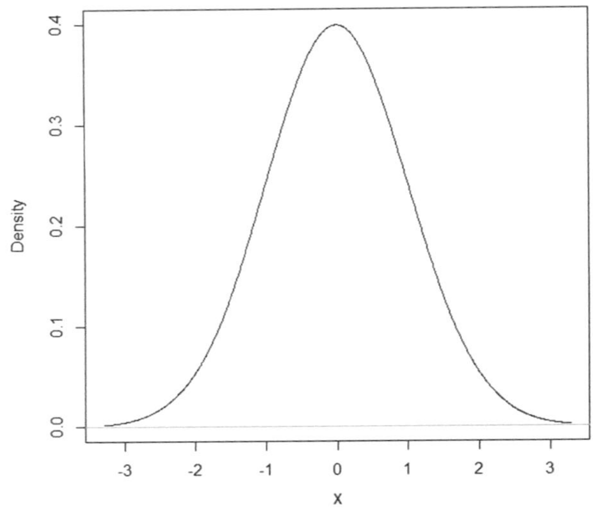

Fig. 3.5 Normal probability density function with mean zero and standard deviation one

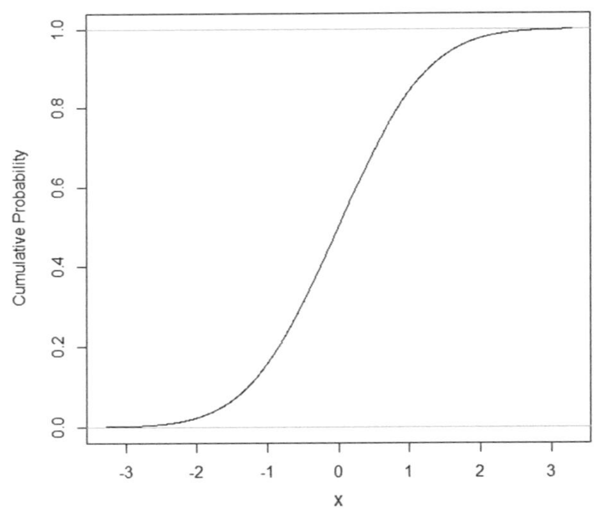

Fig. 3.6 Cumulative normal distribution function with mean zero and standard deviation one

3.5.2 Testing of Hypothesis

In a biological research, we build a theory and would like to support our hypothesis with experimental data. Thus, the testing procedure consists of proposing a hypothesis, assuming the statistical distribution and inferring the conclusion. First, we have an idea that a certain drug is effective against a disease. This is also known as alternative hypothesis. Secondly, we propose a null hypothesis that the drug is not effective against a disease in order to remove the bias. A suitable statistic such as a t-value is computed which is not only a function of random variables but a function

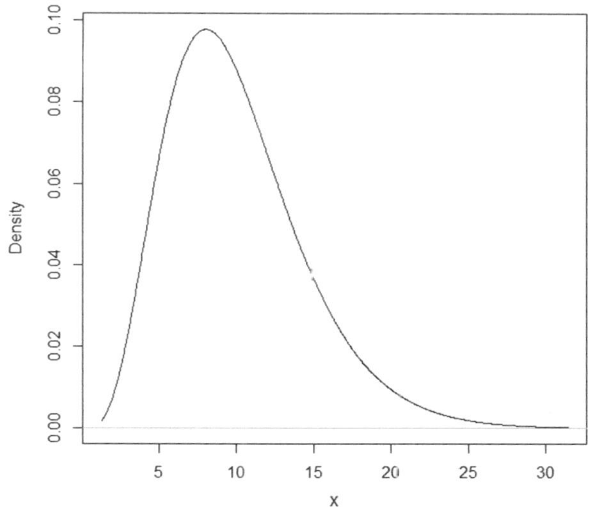

Fig. 3.7 Chi-squared density function with degrees of freedom = 10

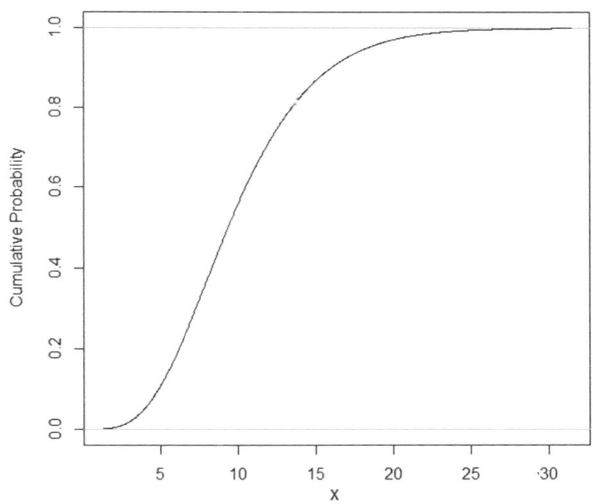

Fig. 3.8 Chi-squared distribution with degrees of freedom = 10

of data values as well. A p-value is computed by comparing the distribution of statistic with the value of statistic. A large p-value indicates that the model is fitting well with the data and the null hypothesis is true with high probability. On the other hand, a low p-value indicates that the experimental outcome is very unlikely under the given assumption of distribution and the null hypothesis is rejected. If the data are found reasonable given the null hypothesis, we accept the null hypothesis and reject our idea. On the other hand, if the data are unreasonable given null hypothesis, the null hypothesis is rejected and idea is provisionally accepted. An acceptance and rejection range of the data are also defined for acceptance and rejection of our idea,

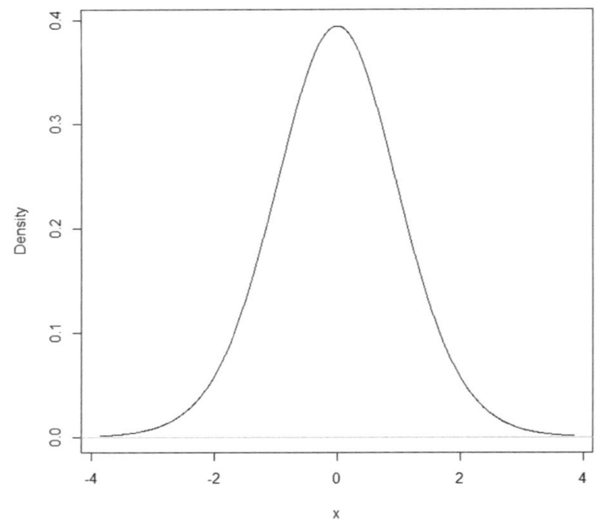

Fig. 3.9 Probability density function of t-distribution with degrees of freedom = 20

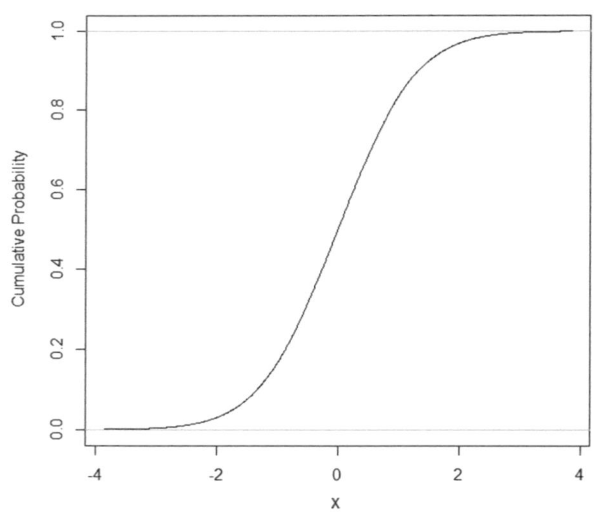

Fig. 3.10 t-distribution with degrees of freedom = 20

respectively. When a null hypothesis is rejected even if it is true, we call it a Type I error. In contrast, a false null hypothesis is accepted in case of Type II error. It is expected that rejection range should be small to minimize Type I errors and acceptance range should also be small to minimize Type II errors. The probability of making a Type I error is known as the significance level of the test. It is generally kept at the 5% level indicating that we are creating Type I error five times in 100 experiments. The power of the test is measured by one minus probability of making Type II error. We always want power of the test to be exceptionally high in case probability of making Type II error is small. If we reduce the significance level

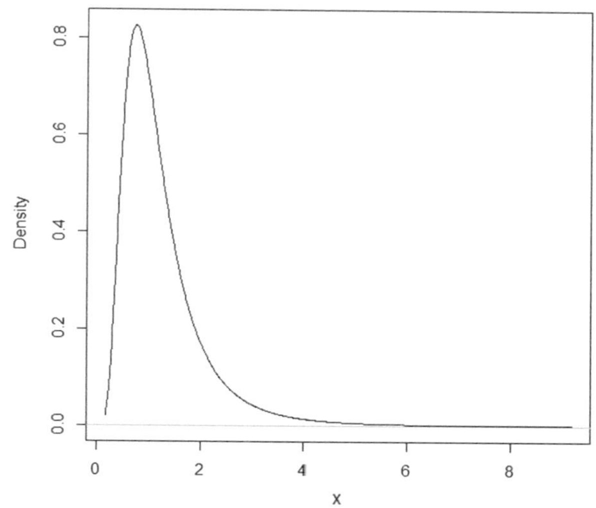

Fig. 3.11 Probability density function of F-distribution with numerator degrees of freedom = 20 and denominator degrees of freedom = 10

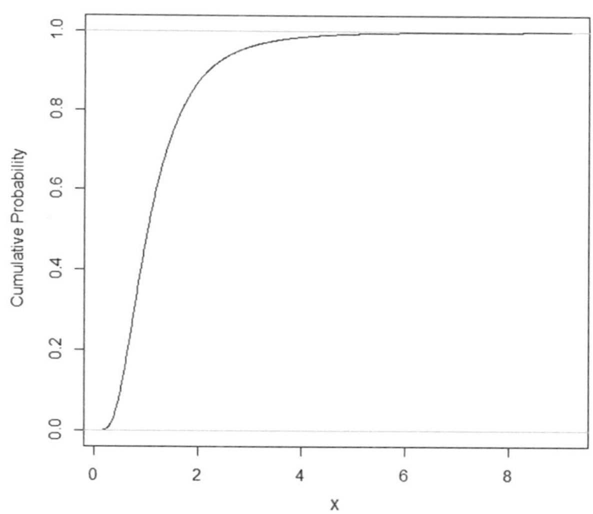

Fig. 3.12 F-distribution with numerator degrees of freedom = 20 and denominator degrees of freedom = 10

by broadening the acceptance range, the power of the test is also reduced making us prone to Type II errors. The power of the test or the probability of drawing correct conclusion can be improved by increasing the sample size prior to conducting experiments.

3.5.3 Parametric Tests Vs. Non-parametric Tests

Parametric tests have some assumptions about the underlying distribution of data. For example, the data collected from a population with normal distribution. The normality of the data is assessed by Shapiro–Wilk test which is based on a Q-Q plot. The Shapiro–Wilk test can be computed in R using built-in function as follows.

```
>shapiro.test (data)
```

Sometimes, parametric tests also assume homogeneity of variance between groups. Most common examples of parametric tests are linear regression, t-tests and analysis of variance. These tests are more powerful than the non-parametric tests.

In case of non-normal distribution due to skewness or heavy tails in the data, rank-based tests are preferred over parametric tests because they do not assume any specific distribution. These tests are also applicable for small sample size.

3.5.4 Common Statistical Tests

3.5.4.1 The Z-Test

This test is applicable in cases where we would like to test null hypothesis against alternative hypothesis and the standard deviation is known. First, the standardized value z is computed from a variable such as gene expression values following a normal distribution. Next, the p-value is defined as the standard normal probability of Z attaining values being more extreme than |z| either in left or right direction. If the p-value is larger than the significance level alpha, then the null hypothesis is not rejected. On the other hand, when the p-value is lesser than the significance level, null hypothesis is eventually rejected.

The R script for computing z-value can be computed as follows.

```
z.value<-sqrt(n)*(mean(x)-mu0)/sigma
p.value<-2*pnorm(-abs(z.value),0,1)
```

Thus, the null hypothesis is rejected when z-value is either highly positive or highly negative. When z-value falls between 95% confidence interval, then null hypothesis is not rejected and consequently this region is known as the acceptance region.

3.5.4.2 One-Sample T-Test

The population standard deviation is unknown in most of the biological research. Therefore, the Z-test cannot be applied under this scenario. The one-sample t-test is an appropriate method under this condition. If the p-value is larger than the significance level, the null hypothesis is not rejected. On the other hand, the null hypothesis is rejected in case p-value is smaller than the significance level. The t-value is computed in R using the following script.

```
>t.value<-sqrt(n)*(mean(x)-mu0)/sd(x)
```

However, there is a built-in function in R which can be easily applied as follows.

```
>t.test(x,mu=0)
```

This one command will generate t-value, p-value and 95% confidence interval for population mean at value 0. The large value of t indicates that the mean is different from 0, whereas very small p-value tending towards zero supports the rejection of null hypothesis.

3.5.4.3 Two-Sample T-Test

One of the most common tests in experimental biology is two-sample t-test. In case, we want to compare the population means of two groups such as healthy individuals and patients. When the variance is unequal between two groups, the test is known as the Welch two-sample t-test. The equal variances between two groups are tested using the built-in R function var.test. A t-statistic is computed based on the mean and the variance of two groups. The computed t-value is large when the sample sizes are large and the standard deviations are small. The t-value is computed in R using the command which is as follows.

```
>t.test (x, y, var.equal= FALSE)
```

If the variances for two groups are equal, the t-value is computed in R using the following script.

```
>t.test(x, y, var.equal = TRUE)
```

Thus, the null hypothesis of equal population means between two groups is rejected if the p-value is less than 0.05.

3.5.4.4 Wilcoxon Rank Test

This two-sample test is widely used in bioinformatics as an alternative to the t-test. We compare the overall distribution of two groups rather than focussing on means. The null hypothesis assumes that both distributions are equal, whereas alternative hypothesis does not support an equal distribution, for example, one distribution is larger or smaller than another one. Here, the data in both groups are ranked and then summed over to compute a statistics W following a correction. The distribution of sum of ranks provides an estimate of p-value which can be used further to reject the null hypothesis in case smaller than the significance level. This test can be applied to a dataset, for example gene expression values between two groups, using the built-in function as follows.

```
>wilcox.test (x,y)
```

3.5.4.5 Chi-Squared Test

This distribution free test is suitable for testing relationship between categorical variables such as gender. The chi-square statistic measures the independence between two variables. The association between two variables is assessed by comparing the observed frequency with the expected frequency in case both variables are independents. The chi-square statistic computed is compared against a critical value under a chi-square distribution. It is also used as a goodness of fit test to compare a sample data with population. A small test statistic indicates that observed value fits well with the expected value and, thus, there is a relationship between two values. On the other hand, a large test statistic indicates that both observed and expected data do not fit well and, therefore, there is no relationship between two types of data. This test can be performed only on numbers not with proportions or percentages, and the number should be large usually more than five. This test can be performed in R using the following script.

```
>chisq.test (data)
```

3.5.4.6 Analysis of Variance (ANOVA)

This test is an alternative to t-test when there are three or more groups under testing. It is common test in bioinformatics and is based on a linear model. We assume that the residuals or errors are normally distributed and have equal variance (homogeneity) across three or more groups. Here, we test a hypothesis that three or more population means are equal. Two sums of squares are important in computation during analysis of variance. Firstly, sum of squares within (SSW) is the sum of the squared deviation of the individual measurements to their group means. Secondly, the sum of squares between (SSB) is the sum of squares of the deviances of the group means in reference to the total mean. Finally, the F-value is computed based on the definition as follows.

$$F = \frac{\mathrm{SSB}\,(G-1)}{\mathrm{SSW}\,(N-G)}$$

where N is the total number of measurements and G is the number of groups under comparison. $G-1$ and $N-G$ are the degrees of freedom. If the data are normally distributed, then the F-value follows F-distribution and the p-value is computed based on this distribution. The value of SSB and F tends to be small under the null hypothesis of equality of group means. If the p-value is equal to or higher than the significance level, then the null hypothesis of equal group means is accepted. In case the p-value is smaller than the significance level, the null hypothesis is rejected. The R function aov() is used to compute one-way ANOVA test and function summary.aov() is used to summarize the ANOVA model. The one-way ANOVA test indicates that some of the group means are different. In order to identify the pairs of groups showing significant difference, Tukey multiple pairwise comparison can be performed using the R function TukeyHSD() where the fitted ANOVA is included as an argument.

3.5.4.7 Kruskal–Wallis Rank Sum Test

This test is a non-parametric alternative to the ANOVA when normality is violated. It is based on ranking of the data and is highly robust against the violation of normality assumption. However, the size of the experimental effects is not estimated in this test. The R function kruskal.test () is used to perform this non-parametric test.

3.5.4.8 Correlation Coefficient

The correlation coefficient is a measure to find linear relationship between two variables. The value of this measure is always between +1 and -1. If the value of the coefficient lies close to either of these two extreme values, then two variables are linearly related. The correlation coefficient in R is computed by the function cor (x, y). The null hypothesis that the correlation coefficient is zero is tested against the alternative hypothesis that correlation coefficient is not zero. Since this test assumes the normal distribution of data, and therefore it provides t-value and corresponding p-value. The correlation between two variables is established by rejecting the null hypothesis of zero correlation if the p-value is very small.

3.5.4.9 Principal Components Analysis

Principal component analysis is a descriptive method to find dependencies between variables using correlations. It finds new directions in the dataset based on the maximal variation. The amount of variance for a particular component is measured by the eigenvalue. Each eigenvalue represents more variance than any other observed variable. The direction of maximal variation is defined as the linear combination with maximal variance. First few variables are important because they have large eigenvalues in comparison to other variables. This property is known as elbow rule. The first principal component is defined as the linear combination of data with first eigenvector. In R, correlations matrix is computed using the built-in function cor followed by computation of eigenvalues and eigenvectors using the built-in function eigen. Alternatively, another built-in function princomp is used to obtain the scores (component scores) and the loadings (eigenvectors).

3.5.5 Graphics in R

R language is useful in creating customized publication quality figures with few commands only. We can draw common plots such as scatter plot, box plot and histogram to compare the distribution of data under different conditions. The R graphics consists of three types of plotting functions: high level, low level and the layout. The high-level plotting involves creation of a complete plot, whereas low-level plotting augments to an existing plot created by a high-level function. Some optional arguments can be included in a high-level function to enhance the quality of the plot. The common arguments are pch for point characters, lty for line type, lwd for line thickness and col for colour. The complete list of colours available in the R is obtained by the command colors (). The x-axis limit and y-axis limit can be fixed using the arguments xlim and ylim, respectively. The arguments xlab and

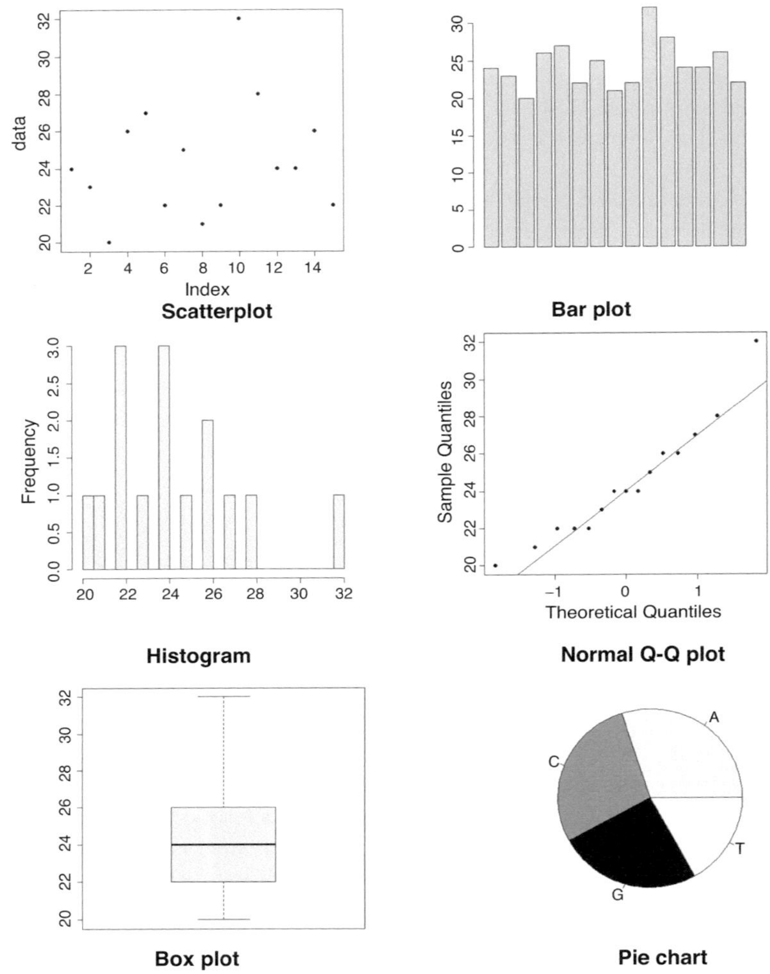

Fig. 3.13 Different plots created using R built-in functions

ylab are included to label the x-axis and y-axis, respectively. Moreover, a command *par* is used to change the default layout of the graphics device. Finally, the plot is exported by saving in common file formats like jpg, postscript, tiff, png and pdf.

Some of the important plots (Fig. 3.13) that can be created using high-level functions are described as follows.

Scatterplot This plot can be used to draw any number of points in a figure window using the command *plot*. It displays number of dots to show the relationship between two variables representing x-axis and y-axis. For example, the expression of genes in disease and healthy conditions can be visualized using scatter plot.

Histogram It is a graphical display of distribution of numerical or categorical data using the command *hist*. It shows the distribution of a dataset into ranges and the height of the bar indicates the number of data included in a particular range. It is usually applied to observe the underlying distribution of data. For example, we can create a histogram to see whether the dataset follows a normal distribution prior to performing a parametric test.

Box Plot A box plot displays the distribution of numerical data based on different levels of quartiles using the command *boxplot*. These levels are minimum, first quartiles (Q1), median, third quartiles (Q3) and maximum. It also provides graphical information not only regarding symmetry and skewness of distribution but indicates the presence of an outlier.

Bar Plot Bar plots are the graphical display of categorical variables in form of rectangles created in R using the *barplot*() function. The height or the length of the bar is proportional to the numerical values.

Normal quantile-quantile (Q-Q) plot The normal quantile plots are the graphical display of the theoretical normal distribution on the x-axis against the dataset on the y-axis to assess whether your dataset deviates from the normal distribution. If the dataset follows normal distribution, we can proceed with the standard statistical analysis using t-test or ANOVA.

Pie Chart The pie chart describes a data graphically in a circular graph. Each slice of a pie represents relative proportion of data. Pie chart is created in R using pie() function.

The low-level functions are only executed on a plot created by high-level functions. For example, *abline* is a low-level function to add a straight line on a scatter plot. Moreover, two functions *lines* and *points* are implemented for adding a line plot and a point to an existing plot, respectively. Similarly, a text on the plot is added by function *text* and on the margin of the plot using a function *mtext*. Finally, a title can be added to an existing plot using the function *title*.

3.5.6 R Packages for Graphics

The base graphics of R has a facility to explore and plot various types of data. Most common package to perform general plotting functions is known as *lattice* developed by Deepayan Sarkar and it is an integral part of the R base installation. It provides a high-level data visualization of multivariate data. The graphical parameters can be customized in *lattice* for a publication quality display. Moreover,

Table 3.1 Commonly used R built-in functions

Category	Function	Description
Input/output functions	getwd()	Find the working directory
	setwd(x)	Set a new working directory
	library (x)	Load a package
	load (x)	Load a saved file
	read.table	Read a space delimited file
	read.csv	Read comma delimited (csv) files
	write.table	Write a tabular data in a file
	x<-c(1,2,3. . .)	Create a data vector x with specified elements
Mathematical functions	sum(x)	Sum of the elements of x
	mean (x)	Mean of the elements of x
	var(x)	Variance of the elements of x
	sd(x)	Standard deviation of elements of x
	cor (x,y)	Linear correlation between x and y
	t(x)	Transpose of x
	%*%	Matrix multiplication
	eigen(x)	Computes eigenvalues and eigenvectors of a matrix
Statistical functions	aov (formula)	Analysis of variance model
	t.test(x,y)	t-test to find significant difference between x and y
	chisq.test	Chi-square test
	var.test	F-test for equal variance
	shapiro.test	Shapiro-Wilk test for normality
	rbinom (n,size,p)	Generate random samples from binomial distribution
	rnorm (n,mean, sd)	Generate random samples from normal distribution
Graphical functions	qqplot	Q-Q plot to check normality of data
	plot (x)	Scatterplot of x
	hist(x)	Histogram of frequencies of x
	barplot(x)	Histogram of values of x
	boxplot(x)	Box and Whiskers plot of x
	pie(x)	Pie chart of x
	abline(a,b)	A straight line of slope b and intercept a

ggplot2 is another powerful package developed by Hadely Wickham to generate publication ready plots. The learning of this package is little challenging but we can expect best return considering the time invested in learning this package. The *qplot()* is a general function in this package that can be implemented to create a wide range of graphs. The options in the graph can be further modified using the *theme()* function. The R built-in functions commonly used are described in Table 3.1.

3.6 Exercises

1. *SRGAP2* and *FOXP2* are the key genes involved in extraordinary cognitive development and speech in human lineage. The expression levels of both genes in the prefrontal cortex of human brain are given below.

SRGAP2: 5.54341, 5.38406, 5.68797, 5.48020, 5.71367, 5.16755, 5.54771, 5.50315, 5.29877, 5.26532, 5.58837, 5.93008, 5.85553, 5.55625.

FOXP2: 2.53967, 3.22162, 2.35884, 2.04029, 2.67516, 2.27110, 1.62437, 2.15451, 1.99223, 2.74845, 2.78851, 2.98586, 2.44189, 2.55143.

1. Compute the mean and standard deviation of gene expression levels of *SRGAP2* and *FOXP2* genes.
2. Perform appropriate tests to check the normality and equality of variances in expression levels of both genes.
3. Perform two-sample t-test to find significant differences in the gene expression levels of both genes.

Solution:

```
>SRGAP2<-c
(5.54341,5.38406,5.68797,5.48020,5.71367,5.16755,5.54771,
5.50315,5.29877,5.26532,5.58837,5.93008,5.85553,5.55625)
>FOXP2<-c
(2.53967,3.22162,2.35884,2.04029,2.67516,2.27110,1.62437,
2.15451,1.99223, 2.74845,2.78851,2.98586,2.44189,2.55143)
> mean(SRGAP2)
[1] 5.537289
> mean(FOXP2)
[1] 2.456709
> sd(SRGAP2)
[1] 0.216255
> sd(FOXP2)
[1] 0.4243888
> shapiro.test(SRGAP2)
 Shapiro-Wilk normality test
data: SRGAP2
W = 0.97621, p-value = 0.9469

  > shapiro.test(FOXP2)
 Shapiro-Wilk normality test
data: FOXP2
W = 0.99442, p-value = 1
> var.test(SRGAP2,FOXP2)
 F test to compare two variances
data: SRGAP2 and FOXP2
```

```
F = 0.25966, num df = 13, denom df = 13, p-value = 0.02122
alternative hypothesis: true ratio of variances is not equal to 1
95 percent confidence interval:
0.0833569 0.8088493
sample estimates:
ratio of variances
 0.2596597
```

The p-value of F-test is 0.02122 which is lower than the significance level 0.05. Therefore, there is a significant difference in between two variances.

```
> t.test(SRGAP2, FOXP2,var.equal=F)
 Welch Two Sample t-test
data: SRGAP2 and FOXP2
t = 24.199, df = 19.325, p-value = 6.436e-16
alternative hypothesis: true difference in means is not equal to 0
95 percent confidence interval:
 2.814441 3.346717
sample estimates:
mean of x mean of y
 5.537289 2.456709
```

2. We want to test a hypothesis that the nucleotides of a gene have equal probability. The null hypothesis to be tested is that the frequency of each nucleotide is 0.25. Compute the total number of nucleotide bases and frequency of each nucleotide in the gene. Perform chi-square test whether nucleotides in the gene occur with equal probability. The coding sequence of the gene is as follows.

ATGAACAGCCAGGGGTCAGCCCAGAAAGCAGGGACACTCCTCCTGT-TGCTGATATCAAAC

Solution

```
> library(Biostrings)
>gene<-DNAString("ATGAACAGCCAGGGGTCAGCCCAGAAAGCAGGGACACT
CCTCCTGTTGCTGATATCAAAC")
> length(gene)
[1] 60
> alphabetFrequency(gene, baseOnly=TRUE, as.prob=TRUE)
 A C G T other
0.3000000 0.2833333 0.2500000 0.1666667 0.0000000
> gene.freq<-alphabetFrequency(gene, baseOnly=TRUE, as.prob=TRUE)
> chisq.test(gene.freq)
 Chi-squared test for given probabilities
```

```
data: gene.freq
X-squared = 0.30278, df = 4, p-value = 0.9896
```

The number of nucleotide bases in the gene is 60 and frequency of A, C, G, and T is 0.3, 0.28, 0.25, and 0.17, respectively. Since, p-value of chi-square test is close to 1, the null hypothesis is not rejected. Thus, the nucleotides of given gene have equal probability.

3.7 Multiple Choice Questions

1. A stored R file can be executed using the command:
   (a) execute
   (b) source
   (c) open
   (d) None of the above
2. The assignment operator in R code is written as:
   (a) <-
   (b) ->
   (c) =
   (d) All the above
3. Which of the following statistical distribution is a discrete distribution:
   (a) Binomial distribution
   (b) Normal distribution
   (c) t-distribution
   (d) F-distribution
4. The command to perform two-sample t-test with unequal variance in R is:
   (a) t.test (x, y, var.equal = TRUE)
   (b) t.test (x, y, var.equal = FALSE)
   (c) t.test(x, y, var.equal = EQUAL)
   (d) t.test (x, y, var.equal = UNEQUAL)
5. Tukey multiple pairwise comparison test is performed in R using command:
   (a) TukeyHSD()
   (b) Tukey.test
   (c) T.test
   (d) TuckeyMPC()
6. The statistical test which is non-parametric alternative to ANOVA test is known as:
   (a) Wilcoxon rank test
   (b) Kruskal–Wallis rank sum test
   (c) Kolmogorov–Smirnov test
   (d) Shapiro–Wilk test

7. High-level plotting in R graphics refers to:
   (a) Creation of a complete plot
   (b) Addition to an existing plot
   (c) Transforming an existing plot
   (d) Modification of an existing plot
8. Which of the following is/are a low-level function in R:
   (a) point
   (b) abline
   (c) text
   (d) All the above
9. The popular R package(s) for graphics is/are:
   (a) lattice
   (b) ggplot2
   (c) lattice and ggplot2
   (d) None of the above
10. Which of the following is a characteristic feature of a data frame?
   (a) Data is numeric, factor or character
   (b) Two-dimensional array-like structure
   (c) Row name is unique and same number of data in each column
   (d) All the above

Answer: 1. b 2. d 3. a 4. b 5. a 6. b 7. a 8. d 9. c 10. d

Summary

- R language is an object-oriented functional programming language used for statistical computing.
- Common data structures in the R environment are vectors, character strings, matrices, lists, data frames and classes.
- R packages are organized libraries for different functionalities available at CRAN and bioconductor.
- Useful statistical parametric and non-parametric tests can be performed in the R environment.
- Both high-level and low-level graphics functions are available in the R environment.

Suggested Reading

Gerrard P, Johnson RM (2015) Mastering scientific computing with R. Packt Publishing, Birmingham

Sinha PP (2014) Bioinformatics with R cookbook. Packt Publishing, Birmingham

Hahne F, Huber W, Gentleman R, Falcon S (2008) Bioconductor case studies. Springer, New York

Lewis PD (2010) R for medicine and biology. Jones and Bartlett Series, Burlington

Wickham H (2016) ggplot2: elegant graphics for data analysis. Springer, Cham

Sequence Alignment

4

Learning Objectives

You will be able to understand the following after reading this chapter:

- Pair-wise and multiple sequence alignment of nucleotide and protein sequences.
- Various scoring matrices for optimal alignment.
- Global and local sequence alignments.

4.1 Introduction

The homologous relationship between different residues in the nucleotide and protein sequences diverged from a common ancestor is inferred using optimal sequence alignment. This idea is based on that fact that highly similar sequences are likely to have same biological functions. During optimal alignment, the homologous residues are arranged in columns either manually or using automated programs in a best possible way even allowing some gaps between the residues. The presence of gaps in the alignment indicates evolutionary insertions or deletions (collectively known as indels) events in the homologous sequences. The sequence alignment is an essential prerequisite computational step before conducting phylogenetic analysis and prediction of protein structure. It is comparatively easy to align closely related sequences and, in turn, the optimal alignment of sequences is an indicator of degree of closeness. However, it is very difficult to perform sequence alignment in case the sequence identity (i.e. number of identical residues/total number of aligned residues) is below 25%. Amino acid alignment is preferred over nucleotide alignment for protein coding sequences because it is easy to perform and less ambiguous. Moreover, the reading frame of the coding nucleotides are prone to break down during

B. K. Tiwary, *Bioinformatics and Computational Biology*,
https://doi.org/10.1007/978-981-16-4241-8_4

nucleotide alignment. The best approach is to first align the sequences at the amino acid level and then reverse translate the residues to corresponding nucleotide sequence alignment. Some programs like DAMBE and TRANSALIGN can perform this kind of analysis. However, it is still very difficult to find optimal alignment of non-coding nucleotide sequences such as introns and repeats.

During alignment, the alignment score for each pair of nucleotides or residues is computed in form of a scoring matrix. This scoring matrix is useful in finding a score of each pair of nucleotides or residues. For example, NCBI-BLAST program assigns a score of +1 for identical nucleotides and − 3 for non-identical nucleotides for highly conserved sequences. The scoring matrix for amino acid sequences are developed considering their observed physicochemical properties and actual substitution frequencies in nature. Point accepted mutation (PAM) matrix is derived from the observed substitution rates. PAM matrices signify the substitution probabilities of amino acids over a constant unit of evolutionary change. For example, PAM-1 is one PAM unit which represents one substitution or accepted point mutation per 100 amino acid residues. This matrix is usually used to compare closely related protein sequences. On the other hand, PAM-1000 matrix may be useful to compare distantly related sequences. Usually, PAM-250 matrix is most commonly used in alignment of protein sequences. BLOSUM matrix is an alternative to the PAM matrix derived from observed substitution rates among related proteins. In BLOSUM matrix, similar protein sequences are first clustered and then substitution rates between clusters are computed. BLOSUM matrices are derived for protein sequences having variable degree of similarity like PAM matrix. However, numbering pattern of BLOSUM matrix is inverse to the numbering pattern of PAM matrix. As an example, a higher number BLOSUM matrix such as BLOSUM-62 is suitable to compare closely related protein sequences, whereas a lower number BLOSUM matrix such as BLOSUM-30 is appropriate for divergent protein sequences having distant relationship. Most commonly used matrices like BLOSUM-62 and BLOSUM-80 are appropriate for comparing protein sequences having approximately 62% and 80% similarities, respectively.

4.2 Pairwise Sequence Alignment

A dot plot is the simplest graphical method for visual comparison of two sequences using a two-dimensional matrix and identify the regions of high similarity. In dot plot, first sequence is plotted on the X-axis, whereas second sequence is plotted on the Y-axis. Presence of identical residues in each position of both sequences is indicated by a dot in the plot space. The adjacent regions with high similarity are indicated by a diagonal in a dot plot. The abundance of dots along a diagonal line is an indicator of multiple identical amino acids in homologous sequences. Thus, this method is useful in finding similarity in large genomic sequences and identifying internal repetitive sequences in protein sequences. However, this method is currently not so popular in bioinformatics analysis and is widely used for educational purpose only.

Sequence 1: ACTGCA
Sequence 2: ACTCA

Match Score = 1
Mismatch Score = -1
Gap Score = -2

A C T G C A
A C T - C A

Score = 3

		A	C	T	C	A
	0	-2	-4	-6	-8	-10
A	-2	1	-1	-3	-5	-7
C	-4	-1	2	0	-2	-4
T	-6	-3	0	3	1	-1
G	-8	-5	-2	1	2	0
C	-10	-7	-4	-1	2	1
A	-12	-9	-6	-3	0	3

Fig. 4.1 Optimum global pairwise alignment using Needleman–Wunsch Algorithm. The optimal traceback path is shown in the scoring matrix table

Dynamic programming is an efficient and fast approach to align homologous sequences with gaps. This algorithm was developed by Needleman and Wunsch (1970) for optimal sequence alignment. The Needleman–Wunsch algorithm breaks down the original problem into smaller sub-problems and then finds an optimal solution. The optimal path starts from beginning of the sequence and terminates at the end of the sequence. Therefore, this alignment is known as global alignment. For example, two sequences ACTGCA and ACTCA are aligned using Needleman–Wunsch algorithm (Fig. 4.1). First, a scoring matrix having dimension of the grids as (m + 1) x (n + 1) is created where m and n are the lengths of the sequences being aligned. The horizontal and vertical axes are labelled with two sequences being aligned. The alignment of two sequences is indicated by a diagonal path from upper left-hand corner to lower right-hand corner. On the other hand, insertion of a gap in the alignment is indicated by a horizontal move and a vertical move in the sequence in the left axis and top axis, respectively. The first row and column of the matrix are filled with multiples of gap penalty. Each grid is assigned a value based on a user-defined match score, mismatch score and gap score. The first grid (2,2) representing first column of the alignment is filled with a score based on one of three possible values: insertion of gap in the first sequence, insertion of gap in the second sequence and alignment of residues. Similarly, the first grid (1,1) is also filled with a score using any of three possibilities. The first possibility is to take a value from the left grid (2,1) and add the gap penalty to obtain the final score indicating insertion of a gap in the sequence along the left axis. The second possibility is to take a value from upper grid (1,2) and add the gap penalty in order to obtain a final score indicating insertion of a gap in the sequence on the top axis. The third alternative is to take a

Sequence 1: ATAATCG
Sequence 2: CGAACG

Match Score = 1
Mismatch Score = -1
Gap Score = -2

```
AA    CG    CG
**    **    **
AA    CG    CG
```

S		C1	G2	A3	A4	C5	G6
	0	0	0	0	0	0	0
A1	0	0	0	1	1	0	0
T2	0	0	0	0	0	0	0
A3	0	0	0	1	1	0	0
A4	0	0	0	1	2	0	0
T5	0	0	0	0	0	1	0
C6	0	1	0	0	0	1	0
G7	0	0	2	0	0	0	2

Score = 2

Fig. 4.2 Local optimal three pairwise alignments using Smith–Waterman algorithm. The lowest score of this matrix is zero and alignment starts from highest score and tracked back to a score of zero. The traceback paths of three alignments are shown in specific colours

value from the grid located diagonally above in the left side (1,1) and add either match or mismatch score to get a final score indicating alignment of residues. The maximum score obtained out of all three possibilities is taken as a final score. Finally, the optimal path of alignment is traced back starting from lower right-hand corner to upper left-hand corner following the alignment score of the matrix. This algorithm finds the best alignment based on the overall alignment score. In order to avoid excessive gaps in the alignment, gaps are usually penalized by subtracting gap penalty from overall score. The gap penalty includes both gap opening penalty and gap extension penalty collectively known as affine gap penalties. Gaps are usually found in blocks in an alignment and a simple linear penalty for gaps is not a method of choice. Instead, affine gap penalties are implemented in many algorithms where gap opening penalty is always higher than the gap extension penalty. Thus, introduction of sequential gaps with affine gap penalty in an alignment has a substantial reduction in the overall cost of penalty.

Smith and Waterman (1981) proposed a modification in the Needleman–Wunsch algorithm in order to allow local alignment (Fig. 4.2). A local alignment finds best matching regions within two sequences being compared. Here, the alignment path does not start from the right-hand corner and end at the left-hand corner. Instead, it can start and end internally anywhere and any direction in the matrix. A zero term is added additionally in the Smith–Waterman algorithm and therefore no value exists in the scoring matrix which is below zero. If a score below zero is obtained by any method, a zero is placed in that position in the matrix. The traceback process begins from the highest value in the matrix representing alignment score and ends after reaching a value of zero.

4.3 Multiple Sequence Alignment

The concept of dynamic programming can be easily extended to alignment of more than two sequences. It finds an alignment which generates best score for the WSP (weighted sum of pairs) objective function. First, a score for each pair of sequences is computed followed by summing over all the pair scores to obtain a WSP function. Here, an extra weight term for each pair is added for more reliable pairs but less weights are assigned to closely related sequences. Overall, this method lacks an evolutionary model and is not suitable for alignment of more than four sequences of moderate length due to high memory and time requirements.

There are some heuristic methods available for progressive multiple sequence alignment based on phylogeny. This method is simple and more attractive and therefore implemented by most of the popular programs. This fast approach can generate a phylogenetic tree from hundreds of sequences followed by their multiple sequence alignment based on the initial tree. First, a neighbour joining phylogenetic tree (also known as guide tree) is generated using pairwise alignment between all sequences and then determining the pairwise distances between them. Subsequently, the sequences are aligned based on the branching order in the tree. First, most closely related sequences are aligned using dynamic programming followed by adding next closely related sequences to the cluster of most closely related sequences in iterative fashion. This iterative process follows the guide tree in placement of the indels among all clustered sequences. This algorithm for progressive alignment has been implemented in most widely used programs such as Clustal W and Clustal X. Clustal W is suitable for alignment of small number of extraordinarily long sequences. However, any misalignment of nucleotides or residues introduced during early alignment phase cannot be corrected during latter phases in the progressive alignment. A better approach to overcome this problem of alignment error is to include consistency-based scoring. This consistency-based scoring is based on WSP function. This approach improves the accuracy of the pairwise alignment by including information from intermediate sequences. Moreover, a new version of Clustal, Clustal Omega generates high-quality alignment of large number of protein sequences based on hidden Markov model. T-COFFEE is a popular program for multiple sequence alignment based on the consistency-based alignment method. However, this program is slower and more computationally demanding than the Clustal W. There are some programs like MUSCLE do iterative refinement towards optimal alignment. MUSCLE can align large number of sequences in a short span of time and, therefore, is implemented in most of the high-throughput pipeline. PROBCONS is another popular program which implements posterior probability-based scoring. Although this method is highly accurate but computationally expensive for large datasets. DIALIGN performs multiple sequence alignment using short conserved regions of similarity. Therefore, it is useful in finding similarity in the conserved isolated domains of small number of sequences. The accuracy of protein sequence alignment can be significantly improved by including structural information. The 3D COFFEE, a variant of T-COFFEE algorithm is available on the webserver EXPRESSO and uses protein structures from PDB database to guide

Table 4.1 Common programs for multiple sequence alignment

Program	Algorithm	URL
CLUSTAL W/ CLUSTAL X	Progressive alignment	www.clustal.org/
CLUSTAL OMEGA	Progressive alignment based on hidden Markov model.	www.clustal.org/
DIALIGN	Greedy and progressive segment-to-segment alignment	dialign.gobics.de/
T-COFFEE	Tree and consistency-based alignment	www.tcoffee.org/
3DCOFFEE	Structure-based alignment	www.tcoffee.org/Projects/expresso/index.html
MAFFT	Iterative refinement and consistency-based alignment	mafft.cbrc.jp/alignment/software/
MUSCLE	Log-expectation score-based progressive alignment and tree-dependent refinement	www.drive5.com/muscle/
PROBALIGN	Maximum expected accuracy alignment based on posterior probability estimation	probalign.njit.edu/standalone.html
PROBCONS	Probabilistic modelling and consistency-based alignment	probcons.stanford.edu/
msa R package	An R package for multiple sequence alignment using Clustal W, Clustal omega, and MUSCLE algorithms	bioconductor.org/packages/release/bioc/html/msa.html

the alignment process. All popular multiple sequence alignment programs are listed in Table 4.1.

4.4 Alignment Visualization and Editing

The multiple sequence alignment generated through automated methods is likely to have some errors. These alignment errors can be removed using manual editing. It is preferable to perform manual editing at the amino acid level for coding sequences. The unambiguously aligned regions of the alignment are expected to be deleted. The decision to remove the gaps from alignment should be taken judiciously. Some gaps contain useful phylogenetic information. It is advisable to retain gaps if they are not inserted in ambiguous manner and alignment columns with gaps are less than 50% of total sites. The presence of gap at a particular site in a sequence may not provide any phylogenetic information about that sequence per se but it does provide useful information regarding its phylogenetic relationship with other aligned sequences. The common software programs available for visualization and manual editing of alignment are listed in Table 4.2. BioEdit is one of the most popular program for visualization and editing of multiple sequence alignment (Fig. 4.3).

Table 4.2 Common programs for visualization and editing of multiple sequence alignment

Editor	Operating system	URL
BIOEDIT	Windows	https://bioedit.software.informer.com
GENEDOC	Windows	https://genedoc.software.informer.com
JALVIEW	Windows, Linux and Mac	https://www.jalview.org/
MACCLADE	Mac	http://macclade.org/macclade.html
SEAVIEW	Windows, Linux and Mac	http://doua.prabi.fr/software/seaview

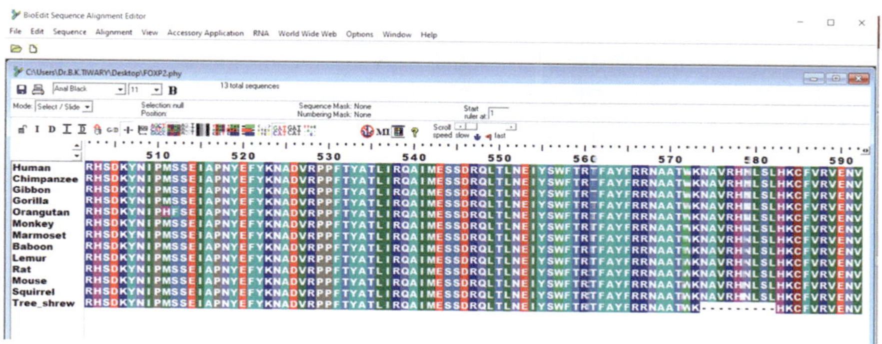

Fig. 4.3 BioEdit sequence alignment editor showing multiple sequence alignment of FOXP2 protein in mammals

4.5 Exercises

1. Find the optimal global alignment and local alignments of two sequences "ATGAAATATACAAGTTATATCTTGGCTTTTCAGCTCTGCATCGTTTTGGGTTCTCTTGGC" and "ATGAACGCTACACACTGCATCTTGGCTTTGCAGCTCTTCCTCA TGGCTGTTTCTGGCTGT" using Biostrings package in the R environment.

Solution

We will use the pairwiseAlignment function in the Biostrings to perform both optimal global and local alignments. The scores for match and mismatch are +1and −1, respectively. The gap opening penalty is 5 and gap extension penalty is 2.

```
>library(Biostrings)
>s1<-DNAString("ATGAAATATACAAGTTATAT")
>s2<-DNAString("ATGAACGCTACACACTGCAT")
>mat <- nucleotideSubstitutionMatrix(match = 1, mismatch = -1,
baseOnly = TRUE)
```

```
>globalAlign <-pairwiseAlignment(s1, s2, substitutionMatrix = mat,
gapOpening = 5, gapExtension = 2)
>localAlign <-pairwiseAlignment(s1, s2, type = "local",
substitutionMatrix = mat,gapOpening = 5, gapExtension = 2)
> globalAlign
Global PairwiseAlignments
pattern: ATGAAATATACAAGT;TATAT
subject: ATGAACGCTACACACTGCAT
score: 4
> localAlign
Local PairwiseAlignments
pattern: [1] ATGAAATATACA
subject: [1] ATGAACGCTACA
score: 6
```

2. Kisspeptin is a recently discovered neuropeptide regulating reproductive function in vertebrates. The protein sequences of human and chimpanzee are retrieved from the Ensembl database and are described as follows:

   Human:
   MNSLVSWQLLLFLCATHFGEPLEKVASVGNSRPTGQQLESLGLLAPGE-QSLPCTERKPAATARLSRRGTSLSPPPESSGSPQQPGLSAPHSRQIPAPQGA-VLVQREKDLPNYNW NSFGLRFGKREAAPGNHGRSAGRG.
   Chimpanzee:
   MNSLVSWQLLLFLCATHFGEPLEEVASVGNSRPTGQQLESLGLLAPGE-QSLPCTERKPAATARLSRRGTSLSPPPESSGSPQPGLSAPNSRQIPAPQGAV-LVQREKDLPNYNWNSFGLRFGKREAAPGNHGRSAGRG.
   Compare the similarity between human and chimpanzee sequences based on a dot plot using seqinr package in the R environment.

Solution Two orthologous sequences are included as a text file "kisspeptin.fasta" in working directory of R environment.

```
>library(seqinr)
> kisspeptin<-read.fasta("kisspeptin.fasta")
> attach(kisspeptin)
> dotPlot(Human, Chimpanzee, wsize = 10, wstep = 6, nmatch = 6)
```

The Fig. 4.4 indicates very high similarities between two neuropeptide sequences from human and chimpanzee as dots are located on the main diagonal in the dot plot.

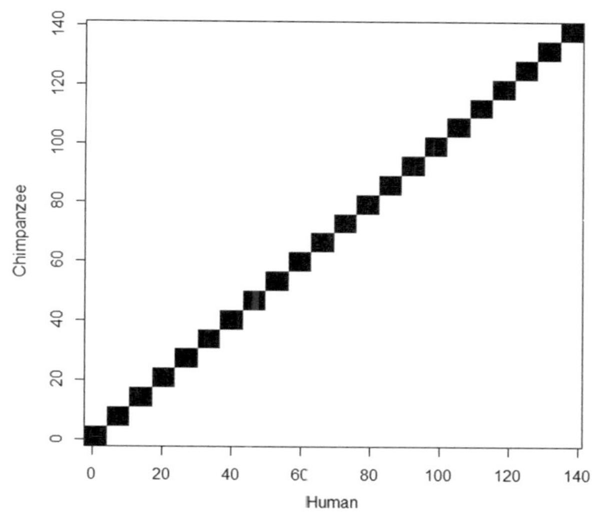

Fig. 4.4 A dot plot showing similarity between human and chimpanzee sequences

4.6 Multiple Choice Questions

1. The presence of a gap in an alignment is due to:
   (a) Insertions only
   (b) Deletions only
   (c) Both insertions and deletions
   (d) None of the above
2. Most appropriate PAM matrix for alignment of closely related sequences is:
   (a) PAM-1
   (b) PAM-250
   (c) PAM-500
   (d) PAM-1000
3. Most commonly used PAM matrix for alignment of protein sequences is:
   (a) PAM-1
   (b) PAM-250
   (c) PAM-500
   (d) PAM-1000
4. Most appropriate BLOSUM matrix for comparing closely related protein sequences having 80% similarity is:
   (a) BLOSUM-20
   (b) BLOSUM-40
   (c) BLOSUM-60
   (d) BLOSUM-80

5. Which of the following statement is correct for dot plot?
   (a) It is a two-dimensional plot
   (b) Identify regions of similarity between two sequences
   (c) Identify internal repetitive regions in protein sequences
   (d) All the above
6. The affine gap penalties consist of:
   (a) Gap opening penalty
   (b) Gap extension penalty
   (c) Gap opening and gap extension penalties
   (d) None of the above
7. In Smith and Waterman algorithm for local alignment, minimum value allowed in a scoring matrix is:
   (a) −1
   (b) 0
   (c) +1
   (d) +2.
8. Which of the following multiple sequence alignment program is based on the protein three-dimensional structures?
   (a) T-COFFEE
   (b) 3D-COFFEE
   (c) PROBCONS
   (d) DIALIGN
9. Most appropriate multiple sequence alignment program for large number of sequences:
   (a) CLUSTAL X
   (b) MUSCLE
   (c) T-COFFEE
   (d) PROBCONS
10. Which of the following program is not used for manual editing and visualization of multiple sequence alignment?
   (a) BIOEDIT
   (b) SEAVIEW
   (c) PROBALIGN
   (d) GENEDOC

Answer: 1. c 2. a 3. b 4. d 5. d 6. c 7. b 8. b 9. b 10. c

Summary

- Amino acid alignment is preferred over nucleotide alignment for protein coding sequences.
- The scoring matrix for amino acid sequences is based on their observed physico-chemical properties and actual substitution frequencies in nature.
- A pair-wise global alignment is performed using Needleman–Wunsch algorithm.
- A pairwise local alignment is performed using Smith and Waterman algorithm.

- The progressive alignment algorithm based on phylogeny is implemented for multiple sequence alignment using Clustal W.

Suggested Reading

Nguyen K, Guo X, Pan Y (2016) Multiple biological sequence alignment: scoring functions, algorithms and evaluation. Wiley, New York

Rosenberg MS (2009) Sequence alignment: methods, models, concepts, and strategies. University of California Press, Berkeley

DeBlasio D, Kececioglu JD (2017) Parameter advising for multiple sequence alignment. Springer, Cham

Chao K-M, Zhang L (2009) Sequence comparison: theory and methods. Springer, London

Structural Bioinformatics

5

Learning Objectives

You will be able to understand the following after reading this chapter:

- Different approaches in computational protein structure modelling such as homology modelling, ab initio modelling and threading.
- Common scoring functions for prediction of protein structure.
- Different approaches in assessment of predicted model.
- Structural classification of protein structures.
- Principles and methods of molecular dynamic simulations.
- Principles and methods of molecular docking.
- Principles and methods of structure-based drug designing.

5.1 Introduction

The structural information regarding the three-dimensional structure of a biological molecule is necessary to understand the mechanism of its function. For example, a detailed understanding of three-dimensional structure of an enzyme is helpful in understanding its mechanism of action in a metabolic reaction. Structural bioinformatics applies various in silico approaches to establish the sequence-structure-function relationship of a biological molecule. In this discipline, diverse computational methods are being developed to predict the protein structure, analyse the protein structure and function and simulate the dynamical behaviour of a protein molecule. Since experimental determination of a protein structure using a high-throughput X-ray crystallography or an NMR spectroscopy method is a tedious and expensive process, computational methods are indispensable in bridging the

B. K. Tiwary, *Bioinformatics and Computational Biology*,
https://doi.org/10.1007/978-981-16-4241-8_5

wide gap between the abundant protein sequence information and protein structure information of millions of protein sequences.

5.2 Protein Structure

A protein structure can be studied at four levels: primary, secondary, tertiary and quaternary structures. The primary structure of a protein consists of a polypeptide sequence of amino acids which starts with the N-terminal amino acid and ends with the C-terminal amino acid. There are 20 standard amino acids present universally among all organisms. The amino acids are linked by a peptide bond which forms an amide linkage between NH_2 group of one amino acid with COOH group of another adjacent amino acid. The peptide bond is unique in terms of having a rigid double-bond character with restricted rotation. The polypeptide backbone forms various conformations known as secondary structures. The alpha helices, beta-pleated sheets and turns or bends are the examples of secondary structures (Fig. 5.1). The secondary structure of a protein further folds in a three-dimensional space to form a tertiary structure. When a protein consists of multiple polypeptides (subunits), each subunit combines together to form a quaternary structure of a functional protein. For

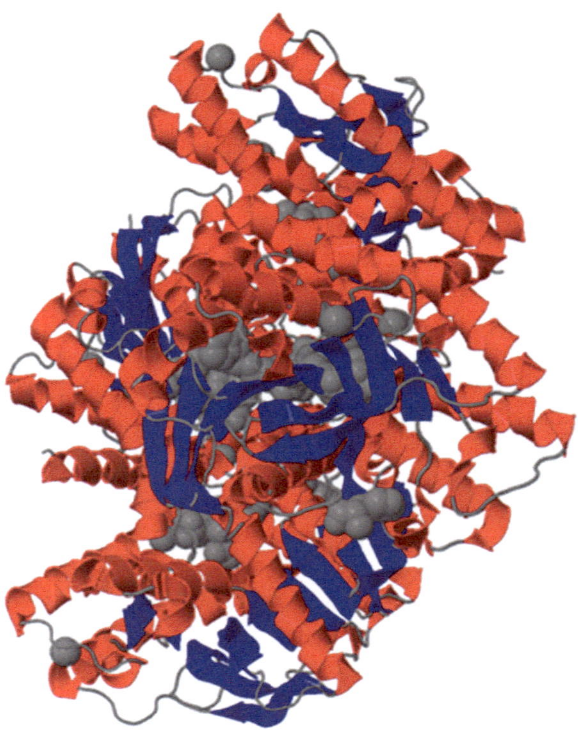

Fig. 5.1 A protein structure of malate dehydrogenase (PDB: 2dfd) showing secondary structures: alpha helices (red) and beta-pleated sheets (blue)

Table 5.1 Molecular graphics applications

Application	Description
RasMol	A popular molecular graphics program written in C
PyMol	A high-performance graphics program written in Java
jmol	An open-source java-based program for molecular graphics
ccp4mg	A comprehensive fully featured program written in Python
VMD	A visual molecular dynamics program for large biomolecules

example, a haemoglobin molecule consists of four functional subunits, two alpha subunits and two beta subunits, forming a functional molecule. Although protein molecules have a definitive three-dimensional structure, many proteins are fully or partially intrinsically disordered under native functional conditions and therefore, are known as intrinsically disordered proteins. Popular molecular graphics applications for visualization of protein structure are listed in Table 5.1.

5.3 Modelling of Protein Structure

The computational modelling of three-dimensional structure of a protein has been one of the grand challenges in bioinformatics. Here, the three-dimensional structure of a protein is predicted from its amino acid sequence. There are four different computational approaches to model the structure of a protein: homology modelling, threading, ab initio modelling (Fig. 5.2) and integrative modelling. Common protein modelling software and severs are listed in Table 5.2.

5.3.1 Homology Modelling

The homology or comparative modelling of a protein is based on the evolutionary principle that protein sequences having common evolutionary descent are likely to have similar three-dimensional structures. Here, available templates with known experimental structures from a related family of proteins are used to model the protein of interest (target protein). First, few appropriate template structures related to the target protein are identified from the Protein Data Bank (PDB) and then correct alignment of template sequence with the target sequence is performed to evaluate the percentage identity between template-target sequence. It is always advisable to prefer sequence identity of 40% or more to attain a good model accuracy where the model deviates less than $2A^0$ RMSD from the template structure. However, we are prone to errors in comparative modelling and the accuracy of the model falls drastically if the sequence identity is below 30%. After alignment, all-atom model of the target protein is generated based on the alignment with one or more available template structures. This all-atom model further undergoes refinement process using an energy minimization step. Finally, the model is evaluated for overall geometrical accuracy of the model based on some specialized scoring methods. We can obtain a

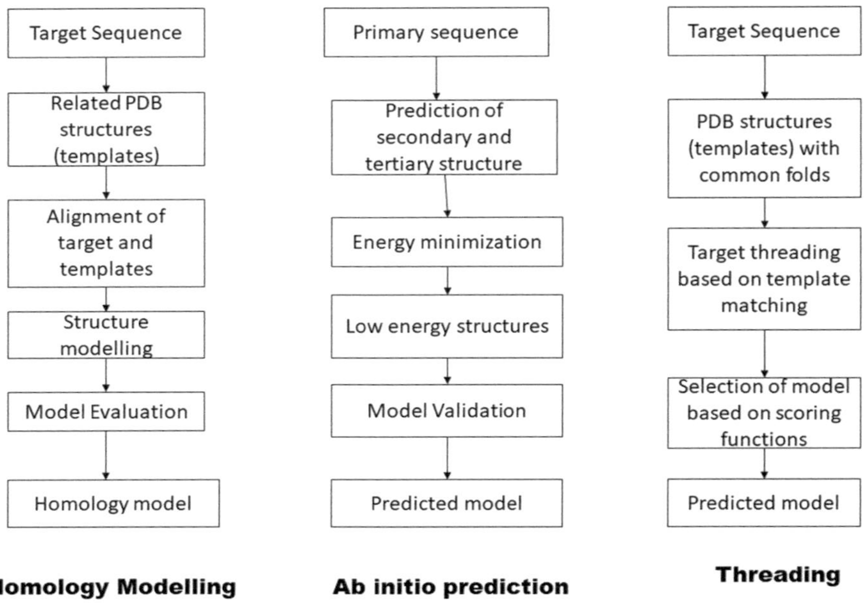

Fig. 5.2 Schematic flowcharts of homology modelling, ab initio prediction and threading

Table 5.2 Protein modelling software and servers

Software/server	Modelling approach	URL
Modeller	Homology modelling	salilab.org/modeller
ModWeb	Homology modelling	salilab.org/modweb
HHPred	Homology modelling	toolkit.tuebingen.mpg.de/tools/hhpred
Rosie	Ab initio prediction	rosie.rosettacommons.org
Robetta	Ab initio prediction	robetta.bakerlab.org
SwissModel	Homology modelling	swissmodel.expasy.org
I-Tasser	Threading	zhanglab.ccmb.med.umich.edu/I-TASSER

very good model with less than $1A^0$ RMSD for the main-chain atoms using homology modelling in case the identity level is above 50%. On the other hand, low accuracy models with multiple types of structural errors are expected with an identity level less than 30%.

5.3.2 Ab Initio Prediction

The structure prediction methods which do not rely on the availability of a template structure are known as de novo methods. Ab initio prediction methods are a subset of de novo methods which depends on energy functions or information inferred from other protein structures. This method assumes a global free energy minimum for the native state of the protein based on the conformational space for available protein tertiary structures with low free energy for a sequence of amino acids. Rosetta method is a popular ab initio method which relies on an ensemble of short structural fragments obtained from the Protein Data Bank (PDB). A Monte Carlo search method is used to assemble the structural fragments based on a scoring function. This method has been applied widely for modelling protein structures and structurally variable regions such as insertions and loops. Another popular method is TASSER which relies on a threading approach to find the local fragments that are subsequently connected into a full-length model by means of a random walk followed by a refinement process. Although de novo methods have predicted accurate models of some small proteins, they suffer from some inherent limitations like huge computational demand and poor quality of models for large proteins.

5.3.3 Threading

The protein fold recognition and threading methods are useful in generating three-dimensional models based on evolutionary distant fold templates. Threading is most popular approach for fold recognition of proteins by fitting a sequence onto the backbone coordinates of available protein structures. Threading has become a generic term for fold recognition although it is a sub-class of fold recognition methods. We can identify the proteins with known structures within the Protein Data Bank (PDB) sharing common folds with the target sequence using fold recognition. These proteins are subsequently used as templates to model the folds of the target sequence. Main focus of this method is to assign folds to the target sequence even if there is low sequence identity to known protein structures. The basic concepts involved in threading is just reverse of the comparative modelling. In comparative modelling, we search for a protein structure best fitting several target sequences. In contrast, each available potential protein structure is fitted with a target sequence in threading process. Here, each target sequence is compared with a library of potential fold templates based on energy potentials and other similarity scores; templates having either least energy score or best similarity score are assumed to best fit of the fold of the target structures.

5.3.4 Integrative (Hybrid) Modelling

Integrative modelling approach computationally combines the structural information from various sources such as X-ray crystallography, NMR spectroscopy, cryo-electron microscopy, cross-linking coupled to mass spectroscopy, sequence analysis, etc. to obtain a structural model of large protein complexes with better precision, accuracy and resolution. Here, the strength of each method complements each other in order to get a better structural model. Integrative modelling iterates through the process of collecting data, proposing an initial model based on the data and again collecting more data for further refinement and validation of the proposed initial model. There are four steps commonly involved in an integrative modelling. First, the input data is collected from various experimental methods and other statistical measures such as atomic statistical potentials and molecular mechanics force fields. A scoring function is developed to evaluate the consistency of the model with the input data. Multiple models are generated and their scores are improved using various optimization algorithms such as Monte Carlo, genetic algorithm and gradient decent. The good scoring models after optimization represent the conformations of the system. Finally, the ensemble of all good scoring models is clustered and analysed to obtain the overall precision and accuracy of the ensemble. It also gives a clue to most probable experiment to be conducted in the next iteration.

5.4 Structural Genomics

The explosive growth of genomic sequences using high-throughput sequencing technology has led to the emergence of a new discipline called functional genomics to determine the function of genes at the genome-wide scale. Structural genomics is the sub-discipline of the functional genomics dealing with structure-based inferences of molecular functions of gene products of all genomic sequence families. This robotics aided large-scale determination of protein structures is known as structural genomics or structural proteomics. This technology-driven field has developed with a fast pace in the last one decade due to recent developments in modern genomic technologies such as cloning and expression of proteins as well as advanced structure determination methods such as macromolecular crystallography and NMR. It provides a systems-level view of protein structural world and further aids in identification of remote homologs which are unlikely to be detected by only sequence comparison.

5.5 Scoring Functions

The three-dimensional structure of a protein is predicted based on the potential energy function. Scoring functions are multidimensional matrices commonly used for structure prediction of proteins. These functions are useful in assessment of both experimentally determined and computationally predicted protein structures. Since scoring functions account for both atom–atom interaction and solvation effect, as a result these functions are also known as effective energy functions. There are four main components of a scoring function: a body definition, a geometrical descriptor, a reference system and a set of restraints. A body definition may consist of either single atoms or centroids of a group of atoms or different atoms or centroids sharing a common physicochemical property. A geometrical descriptor describes interactions between different bodies using some measures such as angles, pairwise distances, dihedral or torsion angles and radial or angular densities. A reference system is the weighted average state of all possible states of a system and used as a reference setting to calculate a specific score for a particular state. A set of restraints either defines the upper and lower limit of the score or divides into different classes requiring a specific treatment.

Some of the most common scoring functions are contact scoring functions, distance-dependent scoring functions, accessible surface scoring functions and combined scoring functions. Contact scoring function is the simplest form of scoring function due to minimum size of the matrix filled with few experimental data. For example, contact potential is a squared two-dimensional matrix of 20 rows and 20 columns for a body definition of alpha carbons of 20 standard amino acids. These functions are generally derived symmetrically for beta carbons or side chain centroids of standard amino acids. Distance-dependent scoring functions are most widely used in the prediction of protein structure. It is represented by a matrix of four dimensions: the first two dimensions include total number of bodies defined, the third dimension distinguishes between local and non-local interactions and the fourth dimension describes distance ranges or bins in order to represent interactions. Accessible surface scoring functions tend to capture the propensity of interaction of defined bodies with the solvent. It describes the solvent accessibility at the residue or atomic level. The combined scoring function is a combination of information from both accessible surface and contact plus distance-dependent approaches. The contact and distance-dependent approach deals with intramolecular protein interactions, whereas accessible surface approach is concerned with the interactions between the solvent and the protein. A normalization step is required to combine these two independent scoring functions.

There are various applications of scoring functions in protein structure prediction. It helps in selecting the best model in terms of accuracy out of different alternative models. Moreover, the fold assessment of a predicted model is performed using a scoring function to evaluate the correct fold. The geometrical accuracy of the overall model as well as individual regions can also be assessed using scoring functions. The stability of a protein structure can also be predicted using scoring functions in case of mutant screening.

5.6 Model Assessment

The primary goal of model assessment is to evaluate the overall accuracy of the predicted model. In addition, it also selects the most accurate model from a set of alternative models and evaluates the accuracy of the different regions of the model. There are four common approaches of model assessment, namely physics-based energies, knowledge-based potentials, combined scoring functions and clustering approaches. Chemical force fields are the physics-based energy function of particles in a system and are derived from both quantum mechanical calculations and experimental findings. The force field energy function is represented by the energy from chemical bonds between interacting atoms (i.e. distances, angles and dihedral angles) and interactions between non-bonded atoms (i.e. electrostatic and van der Waals interactions). These physics-based approaches are useful in selecting near-native structure models from a set of predicted models. The second approach is the knowledge-based potential based on empirical observations on the known protein structures. It is implemented in form of statistical potentials or potentials of mean force. Several statistical potentials such as contact, distance, solvent accessibility and pairwise interaction along with solvent accessibility are derived to assess the structural features of a model. The third approach is combined scoring functions which are an optimized weighted combination of various scores from physics and knowledge-based approaches. However, all three scoring approaches often fail to detect the near-native structure. Finally, multiple models are generated from the same sequence and are compared structurally to select the best model in the clustering approach. The quality of the scoring functions used to generate an ensemble of conformations determines the accuracy of the model. Thus, useful information from an ensemble of possible conformations are converted to statistical probabilities in order to identify the near-native structure of the protein. However, clustering approach cannot assess the quality of the model alone and aided by other scoring functions.

5.7 Classification of Protein Structures

The structural classification of proteins is largely based on the identification of individual domains and their evolutionary relationships. The structural similarity between two proteins is usually assessed by superimposing the C alpha atoms of a protein structure on the top of another one. The distances between two equivalent atoms in both superimposed proteins are measured in terms of root mean square deviation (RMSD). A low RMSD value less than 3.5 A^0 indicates a close similarity in protein structures and consequently a possible structural homology between two proteins. There are only two manually curated protein structure classification databases, namely CATH database and SCOP database covering the entire protein structures deposited in the PDB. The CATH stands for four levels of classification hierarchy, namely Class, Architecture, Homologous superfamily and Topological motif, and is regularly updated using automated algorithms followed by manual curation. Domains are classified in the CATH database using both automated and manual approaches based on similarity at the sequence, structure and functional levels. Firstly, domains are classified into three main classes based on the secondary structures: mainly alpha-helical structures, mainly beta sheet structures and mixed alpha-beta structures. Further, they are classified into 40 groups based on the architecture or gross arrangement of secondary structures independent of their connectivity in a three-dimensional space. Thirdly, they are classified into more than 1000 topological motifs or fold groups based on the arrangement of secondary structures and their connectivity as well. Finally, domains are clustered into more than 2000 homologous superfamilies based on their evolutionary relationships. These domains also share substantial structural similarity along with similarity at the sequence and/or functional level as well.

The structural classification of proteins (SCOP) database also divides a protein structure into one or more domains like the CATH database but the entire classification process is manual. It follows a classification hierarchy at various major levels. The highest level is the class which is arranged according to the secondary structure content of the protein. The SCOP contains five major classes such as all alpha, all beta, alpha/beta (interspersed alpha helices and beta strands), alpha+beta (segregated alpha helices and beta strands) and multidomain (having two or more domains from different classes). The architecture level is missing in the SCOP database unlike CATH database. However, fold groups and superfamily groups are included in the hierarchy of SCOP database like CATH database.

5.8 Molecular Dynamics Simulations

Molecular dynamics simulations are linked to the concept of intermolecular interactions in a complex system. These models are based on the force field and provide useful insights into various intermolecular processes such as protein folding, protein-ligand interactions and protein dynamics. Molecular dynamics uses simple approximations to the laws of Newtonian physics in order to simulate atomic motions (atomistic representation). The process of molecular dynamic simulation can be performed in a few steps (Fig. 5.3). The initial model of the molecular system is obtained from the experimental structure derived from nuclear magnetic resonance (NMR) and X-ray crystallography or computational homology modelling. The forces operating on each atom of the system are estimated from a potential energy function equation describing the forces derived from interactions between bonded and non-bonded atoms which are as follows:

$$\text{Total energy function } E_{\text{total}} = [E_b + E_\theta + E_\phi + E_\omega] + [E_{\text{vdW}} + E_{\text{elc}}]$$

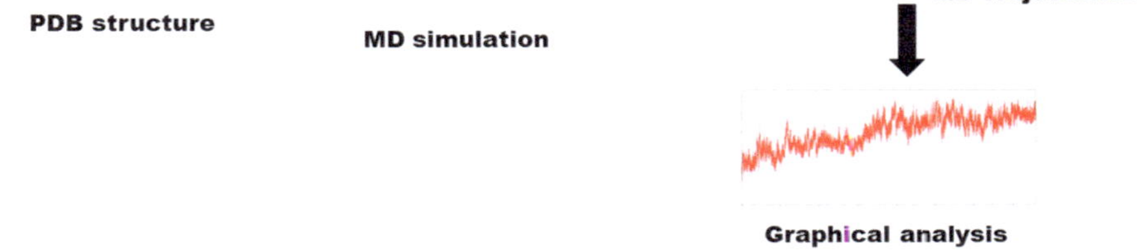

Fig. 5.3 The workflow of a molecular dynamic (MD) simulation of a PDB structure (PDB:2dfd) generating a large dataset of trajectories for further graphical analysis

Table 5.3 MD simulation software

Program	Description
GROMOS	A package for dynamic modelling of biomolecules
GROMACS	A package for high-performance molecular dynamics of biomolecules
NAMD	A parallel molecular dynamics program for high-performance simulation.
AMBER	A package of biomolecular simulation programs
CHARMM	A molecular simulation program for biological systems
Bio3d	An R package to analyse protein structure and trajectory data
VMD	A software to analyse and animate trajectories of MD simulation
MDplot	An R package providing plotting functions to MD simulation output

where E_b is a bond length potential term, E_θ is a bond angle potential term, E_φ is a dihedral angle (torsion) potential term, E_{vdW} is a non-bonded van der Waals interaction term and E_{elc} is a non-bonded electrostatic term. In addition to these energy functions, some constraints are often applied to force a molecule in a desired conformation as an optional term of the total potential energy function.

Chemical bonds and atomic angles are modelled based on simple virtual springs (bond stretching and angle bending), whereas dihedral angles (rotations about a bond) are modelled using a sinusoidal function. Non-bonded forces such as van der Waals interactions and electrostatic interactions are modelled using Lennard-Jones potential and Coulomb's law, respectively. The actual behaviour of real molecules in motion can only be modelled after parameterizing the energy terms to fit quantum mechanical calculations. The parameterizing process consists of identification of ideal stiffness and length of the spring representing chemical bonds and atomic angles, estimation of best partial atomic charges for calculating electrostatic interaction energies, identification of proper van der Waals atomic radii, and so forth. All these parameters are known as force field due to their individual contributions of atomic forces controlling molecular dynamics. There are some useful engines for molecular dynamic simulation (Table 5.3). The GROMOS (GROningen MOlecular Simulation), AMBER (Assisted Model Building with Energy Refinement), CHARMM (Chemistry at HARvard Molecular Mechanics) and GROMACS (GROningen MAchine for Chemical Simulations) are commonly used force fields that only differ in their mode of parameterization. After estimation of the forces acting on the system atoms, the position of atoms is moved following the Newton's laws of motion. Further, simulation time is repeatedly advanced by a small fraction of a second for millions of times. These simulations are usually performed on computer clusters or supercomputers due to large number of calculations required. The MD simulation output are analysed using standard plots such as root mean square deviation (RMSD) and root mean square fluctuation (RMSF). RMSD is the measure of positional divergence of atoms during simulation over a time range (Fig. 5.4), whereas RMSF indicates the degree of positional variation of a certain atom over time (Fig. 5.5). The major limitations of molecular dynamics simulations are requirement of further refinement of force fields and very high computational demand.

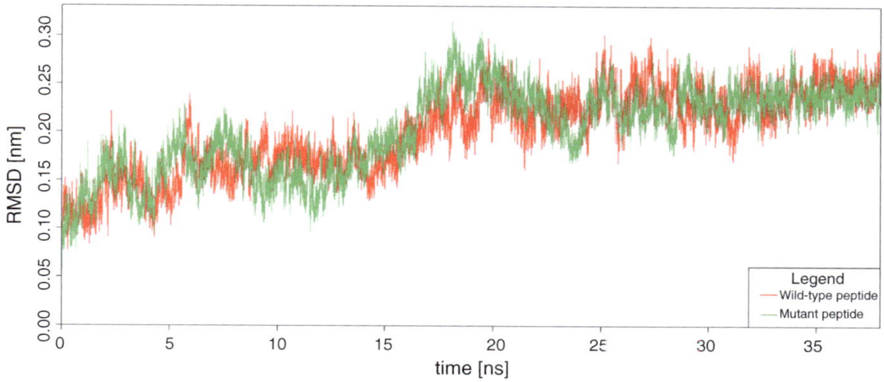

Fig. 5.4 The RMSD curves of two independent MD simulation trajectories in 35 nanoseconds

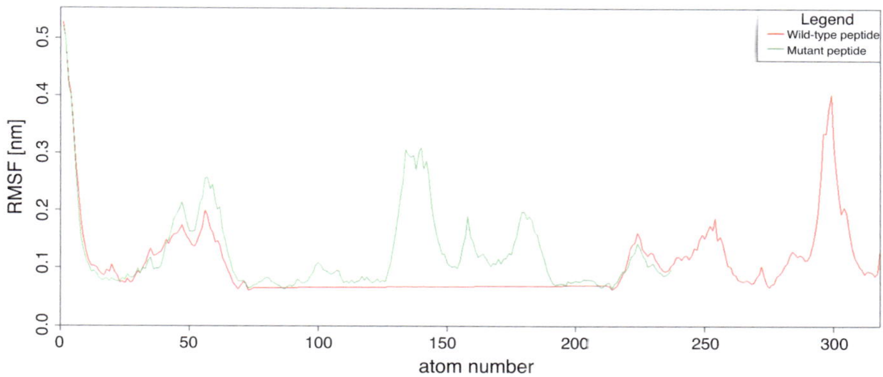

Fig. 5.5 The RMSF curves of two independent MD simulations showing positional variation of first 300 atoms

5.9 Docking of Ligands and Proteins

The interaction between small-molecule ligands or proteins with their protein receptors in an aqueous solution is crucial in understanding the mechanistic basis of pharmaceutically active compounds. The ligand-receptor complex is not only stabilized by the intermolecular van der Waals interactions, hydrogen bonds and electrostatic interactions but also by the desolvating nonpolar parts of the molecular surface. Both ligand and receptor interact with solvent before binding. The desolvation of the nonpolar parts of the surface releases water molecules resulting in the increase of entropy which further strengthens complex formation. This interaction between the ligand and its receptor is known as docking (Fig. 5.6) and has laid a foundation for the computational structure-based drug design. Some

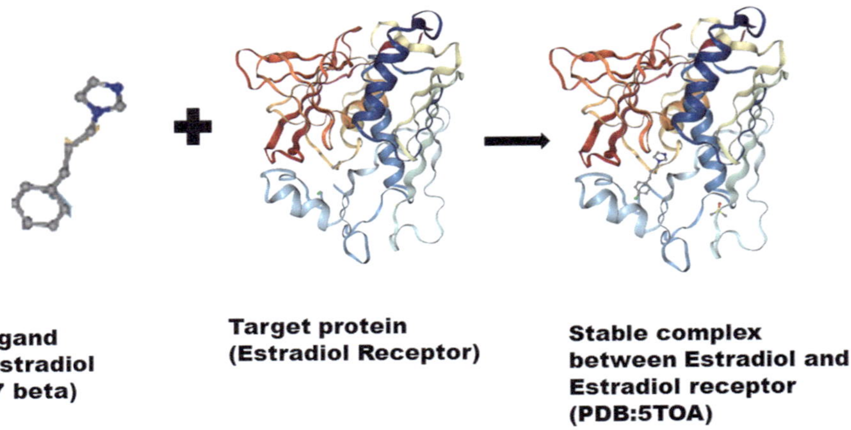

Fig. 5.6 Docking of a small ligand into the crystal structure of its receptor to form a stable complex (PDB:5toa)

popular docking programs such as Autodock, Glide, GOLD and LigandFit have been developed on the fundamental principles of physical chemistry. Molecular docking is useful in finding and optimizing lead compounds and thereby proved as a powerful tool of drug discovery. A docking program has two important components: a docking algorithm to search the configurational and conformational degrees of freedom and a scoring function for evaluation. The docking algorithm tends to find the global energy minimum by searching the potential energy landscape extensively. Only translational and rotational degrees of freedom are allowed for the ligand interacting with receptor active site in rigid docking. In contrast, flexible ligand docking involves torsional degrees of freedom for the ligand by allowing variation in ligand dihedral angles. The scoring function, in general, evaluates not only the steric complementarity between the ligand and its receptor but their chemical complementarity as well. A popular docking program Autodock implements simulated annealing Monte Carlo algorithms having tens of thousands of steps during each cycle. First, the ligand is placed randomly onto the binding site of the receptor and allowed to move towards a global energy minimum. The structure is minimized after each move and subsequently the energy of the new structure is measured. The simulation may consist of several cycles with decreasing temperature in each cycle. The lowest energy in the previous cycle becomes the starting point of the next cycle.

The protein-ligand binding affinity is measured in terms of free energy gains. The largest contributor to this free energy gain emerges from the displacement of water from the hydrophobic regions of the receptor. Another important source of free energy after ligand binding is the hydrogen bonds between the ligand and receptor protein. This free energy gain is achieved by displacing the water bound to the receptor protein by the ligand. The purpose of a docking program is to use a global scoring function that can be equally applied to all protein-ligand complexes. The scoring function, in general, consists of three force field terms for van der Waals interactions, hydrogen bonding and electrostatic interactions and a term representing for desolvation effects. The strength of interaction between a ligand and a protein in terms of binding free energy ΔG_{bind} is as follows.

$$\Delta G_{\text{bind}} = E_{\text{MM}} - T\Delta S_{\text{Solute}} + \Delta G_{\text{Sclvent}}$$

where E_{MM} is the internal energy of the molecule determined from the bond lengths, the bond angles and the dihedral angles followed by enthalpic contributions to free energy from the van der Waals interactions and the electrostatic interactions. The solute entropy consists of terms for translational, rotational, vibrational and conformational entropy. The solvent free energy includes both polar and nonpolar terms and represents both entropic and enthalpic effects of the solvent. The accuracy of a docking program is usually evaluated by comparison of the predicted ligand-receptor complex structure with the experimental structure obtained by X-ray crystallography. During such comparison, each ligand is docked to its own cognate receptor conformation (self-docking). Sometimes accuracy of the program is also tested by docking a ligand to non-cognate receptor conformations such as apo structure of receptor or co-crystallized structure with a different ligand. This method of testing accuracy of a docking program is known as cross-docking.

5.10 Structure-Based Drug Design

A small-molecule ligand is tailored to fit the three-dimensional topology of the binding pocket of a target protein during structure-based drug design. This computational method has a very important role in the overall drug discovery process. The availability of a three-dimensional structure of target protein at atomic resolution preferably with a bound ligand is prerequisite for a structure-based drug design. The first and foremost challenge of this process is to quantify the binding affinity of a ligand-protein complex. Quantitative structure-activity relationships (QSAR) is the most common and an efficient approach to determine quantitatively the binding affinity of a complex using linear or multiple regression methods. This technique is widely used not only for lead identification and optimization but also to predict ADMET (adsorption, distribution, metabolism, elimination, toxicity) properties in pharmaceutical industries. QSARs attempt to find a correlation between the structural properties of potential drug candidates and their binding affinity towards a macromolecular target. The structure-based drug design has become more powerful with ever increasing availability of three-dimensional structures from X-ray diffraction analysis. The structure-activity relationships may also depend upon the three-dimensional structure of a ligand (3D-QSAR) using comparative molecular field analysis (CoMFA). Here, a grid or a surface is used as a surrogate of the binding site of the protein receptor (pharmacophore) having all its steric and electrostatic properties. A series of compounds in their bioactive conformations are then superimposed around this surface or grid. The quality of this predictive model depends upon the correct superposition of ligands and available structural information regarding the target protein. The 4D-QSAR is an advanced form of multidimensional QSAR (mQSAR) where energetically feasible binding modes are included in a 4D-dataset such as different ligand conformations, ligand poses and protonation states. Using this dataset, the actual bioactive conformation or true binding mode of the ligand is identified based on the QSAR algorithm. Each ligand molecule is represented as a single entity in the 3D-QSAR, whereas the ligand molecule is represented in an ensemble of conformations, poses, protonation states, tautomeric forms and stereoisomers in the 4D-QSAR. The binding pocket of a protein structure is not only modified by a bound ligand (induced-fit effect) but other biochemical properties of a receptor protein such as hydrophobicity, hydrophilicity and solvent accessibility are also altered. In some proteins like GPCR proteins, the structure of target protein is not known and realistic induced-fit effect cannot be determined. Thus, a fifth dimension has been added in the 5D-QSAR where we can identify realist induced-fit protocols in addition to 3D topology of the binding pocket. The protein target is strongly affected by solvation phenomena (ligand desolvation, solvent stripping and proton transfer) when bound to a small-molecule ligand. Thus, 6D-QSAR accounts for different solvation models simultaneously and thereby allows more realistic simulation of the ligand-protein binding process. 6D-QSAR is implemented in a receptor modelling program known as *Quasar* allowing optimization based on genetic algorithm. The binding energy in *Quasar* is determined as:

$$E_{\text{binding}} = E_{\text{ligand-receptor}} - E_{\text{ligand desolvation}} - E_{\text{ligand strain}} - T\Delta S - E_{\text{induced fit}}$$

where $E_{\text{ligand-receptor}}$ is composed of $E_{\text{electrostatic}}$, $E_{\text{van der Waals}}$, $E_{\text{hydrogen bonding}}$ and $E_{\text{polarization}}$.

Raptor is another popular program for 6D-QSAR based on a different scoring function. The underlying scoring function is composed of directional terms for hydrogen bonding and hydrophobicity considering solvation effect implicitly. The binding energy in Raptor is calculated as:

$$E_{\text{binding}} = E_{\text{ligand-receptor}} - T\Delta S - E_{\text{induced fit}}$$

where E ligand-receptor is composed of $E_{\text{hydrogen bonding}}$ (shell 1), $E_{\text{hydrophobic}}$ (shell 1), $E_{\text{hydrogen bonding}}$ (shell 2) and $E_{\text{hydrophobic}}$ (shell 2).

5.11 Exercises

1. Malate dehydrogenase is homodimer enzyme catalysing conversion of oxaloacetate and malate in citric acid cycle. Read the human malate dehydrogenase PDB structure (ID:2DFD) in the R environment and find the total number of chains, atoms and residues present in the structure. Compute the binding sites and plot the B-factors of residues present in the enzyme.

Solution (Fig. 5.7)

We will load an R package "bio3d" for reading and finding different properties of PDB structure using following scripts:

```
> library(bio3d)
> help(package=bio3d)
> MDH<-read.pdb("2dfd")
> MDH
 Call: read.pdb(file = "2dfd")
Total Models#: 1
Total Atoms#: 10213, XYZs#: 30639 Chains#: 4 (values: A B C D)
Protein Atoms#: 9195 (residues/Calpha atoms#: 1260)
Nucleic acid Atoms#: 0 (residues/phosphate atoms#: 0)
 Non-protein/nucleic Atoms#: 1018 (residues: 814)
Non-protein/nucleic resid values: [ CL (7), HOH (799), MLT (4), NAD (4) ]
> bs<-binding.site(MDH)
> bs
$inds
Call: NULL
Atom Indices#: 1310 ($atom)
XYZ Indices#: 3930 ($xyz)
+ attr: atom, xyz
$resnames
 [1] "LEU 12 (A)" "GLY 13 (A)" "SER 15 (A)" "GLY 16 (A)" "GLY 17 (A)"
 [6] "ILE 18 (A)" "GLY 19 (A)" "TYR 38 (A)" "ASP 39 (A)" "ILE 40 (A)"
 [11] "ALA 41 (A)" "PRO 81 (A)" "ALA 82 (A)" "GLY 83 (A)" "VAL 84 (A)"
 [16] "PRO 85 (A)" "ARG 86 (A)" "THR 91 (A)" "ARG 92 (A)" "ASP 93 (A)"
 [21] "LEU 95 (A)" "ASN 99 (A)" "ILE 102 (A)" "LEU 106 (A)" "ILE 122 (A)"
 [26] "ALA 123 (A)" "ASN 124 (A)" "VAL 126 (A)" "PRO 145 (A)" "ASN 146 (A)"
 [31] "VAL 151 (A)" "LEU 154 (A)" "ASP 155 (A)" "ARG 158 (A)" "HIS 182 (A)"
 [36] "ALA 183 (A)" "GLY 184 (A)" "GLY 216 (A)" "THR 217 (A)" "VAL 220 (A)"
 [41] "SER 228 (A)" "ALA 229 (A)" "THR 230 (A)" "MET 233 (A)" "LYS 261 (A)"
 [46] "ILE 281 (A)" "LEU 12 (B)" "GLY 13 (B)" "SER 15 (B)" "GLY 16 (B)"
 [51] "GLY 17 (B)" "ILE 18 (B)" "GLY 19 (B)" "TYR 38 (B)" "ASP 39 (B)"
 [56] "ILE 40 (B)" "ALA 41 (B)" "PRO 81 (B)" "ALA 82 (B)" "GLY 83 (B)"
 [61] "VAL 84 (B)" "PRO 85 (B)" "ARG 86 (B)" "ARG 92 (B)" "LEU 95 (B)"
 [66] "ASN 99 (B)" "ILE 102 (B)" "LEU 106 (B)" "ILE 122 (B)" "ALA 123 (B)"
 [71] "ASN 124 (B)" "VAL 126 (B)" "VAL 151 (B)" "LEU 154 (B)" "ASP 155 (B)"
 [76] "ARG 158 (B)" "HIS 182 (B)" "ILE 191 (B)" "VAL 198 (B)" "ASP 199 (B)"
 [81] "PHE 200 (B)" "LEU 205 (B)" "GLY 216 (B)" "THR 217 (B)" "VAL 220 (B)"
 [86] "SER 228 (B)" "ALA 229 (B)" "THR 230 (B)" "MET 233 (B)" "LEU 12 (C)"
 [91] "GLY 13 (C)" "SER 15 (C)" "GLY 16 (C)" "GLY 17 (C)" "ILE 18 (C)"
 [96] "GLY 19 (C)" "TYR 38 (C)" "ASP 39 (C)" "ILE 40 (C)" "ALA 41 (C)"
[101] "PRO 81 (C)" "ALA 82 (C)" "GLY 83 (C)" "VAL 84 (C)" "PRO 85 (C)"
[106] "ARG 86 (C)" "THR 91 (C)" "ARG 92 (C)" "ASP 93 (C)" "LEU 95 (C)"
```

```
[111] "ASN 99 (C)" "ILE 102 (C)" "THR 105 (C)" "LEU 106 (C)" "ILE 122 (C)"
[116] "ALA 123 (C)" "ASN 124 (C)" "VAL 126 (C)" "VAL 151 (C)" "LEU 154 (C)"
[121] "ASP 155 (C)" "ARG 158 (C)" "HIS 182 (C)" "ALA 183 (C)" "GLY 184 (C)"
[126] "GLY 216 (C)" "THR 217 (C)" "VAL 220 (C)" "SER 228 (C)" "ALA 229 (C)"
[131] "THR 230 (C)" "MET 233 (C)" "LYS 261 (C)" "SER 262 (C)" "GLN 263 (C)"
[136] "LEU 12 (D)" "GLY 13 (D)" "SER 15 (D)" "GLY 16 (D)" "GLY 17 (D)"
[141] "ILE 18 (D)" "GLY 19 (D)" "TYR 38 (D)" "ASP 39 (D)" "ILE 40 (D)"
[146] "ALA 41 (D)" "PRO 81 (D)" "ALA 82 (D)" "GLY 83 (D)" "VAL 84 (D)"
[151] "PRO 85 (D)" "ARG 86 (D)" "THR 91 (D)" "ARG 92 (D)" "ASP 93 (D)"
[156] "LEU 95 (D)" "ASN 99 (D)" "ILE 102 (D)" "LEU 106 (D)" "VAL 121 (D)"
[161] "ILE 122 (D)" "ALA 123 (D)" "ASN 124 (D)" "VAL 126 (D)" "VAL 151 (D)"
[166] "LEU 154 (D)" "ASP 155 (D)" "ARG 158 (D)" "HIS 182 (D)" "ALA 183 (D)"
[171] "GLY 184 (D)" "ILE 191 (D)" "VAL 198 (D)" "ASP 199 (D)" "PHE 200 (D)"
[176] "LEU 205 (D)" "GLY 216 (D)" "THR 217 (D)" "VAL 220 (D)" "SER 228 (D)"
[181] "ALA 229 (D)" "THR 230 (D)" "MET 233 (D)"
$resno
 [1] 12 13 15 16 17 18 19 38 39 40 41 81 82 83 84 85 86 91
 [19] 92 93 95 99 102 106 122 123 124 126 145 146 151 154 155 158 182 183
 [37] 184 216 217 220 228 229 230 233 261 281 12 13 15 16 17 18 19 38
 [55] 39 40 41 81 82 83 84 85 86 92 95 99 102 106 122 123 124 126
 [73] 151 154 155 158 182 191 198 199 200 205 216 217 220 228 229 230 233 12
 [91] 13 15 16 17 18 19 38 39 40 41 81 82 83 84 85 86 91 92
[109] 93 95 99 102 105 106 122 123 124 126 151 154 155 158 182 183 184 216
[127] 217 220 228 229 230 233 261 262 263 12 13 15 16 17 18 19 38 39
[145] 40 41 81 82 83 84 85 86 91 92 93 95 99 102 106 121 122 123
[163] 124 126 151 154 155 158 182 183 184 191 198 199 200 205 216 217 220 228
[181] 229 230 233
$chain
 [1] "A" "A" "A" "A" "A" "A" "A" "A" "A" "A" "A" "A" "A" "A" "A" "A" "A" "A"
 [19] "A" "A" "A" "A" "A" "A" "A" "A" "A" "A" "A" "A" "A" "A" "A" "A" "A" "A"
 [37] "A" "A" "A" "A" "A" "A" "A" "A" "A" "A" "B" "B" "B" "B" "B" "B" "B" "B"
 [55] "B" "B" "B" "B" "B" "B" "B" "B" "B" "B" "B" "B" "B" "B" "B" "B" "B" "B"
 [73] "B" "B" "B" "B" "B" "B" "B" "B" "B" "B" "B" "B" "B" "B" "B" "B" "B" "C"
 [91] "C" "C" "C" "C" "C" "C" "C" "C" "C" "C" "C" "C" "C" "C" "C" "C" "C" "C"
[109] "C" "C" "C" "C" "C" "C" "C" "C" "C" "C" "C" "C" "C" "C" "C" "C" "C" "C"
[127] "C" "C" "C" "C" "C" "C" "C" "C" "C" "D" "D" "D" "D" "D" "D" "D" "D" "D"
[145] "D" "D" "D" "D" "D" "D" "D" "D" "D" "D" "D" "D" "D" "D" "D" "D" "D" "D"
[163] "D" "D" "D" "D" "D" "D" "D" "D" "D" "D" "D" "D" "D" "D" "D" "D" "D" "D"
[181] "D" "D" "D"
> plot.bio3d(MDH$atom$b[MDH$calpha], sse=MDH, typ="l", ylab="B-
factor")
```

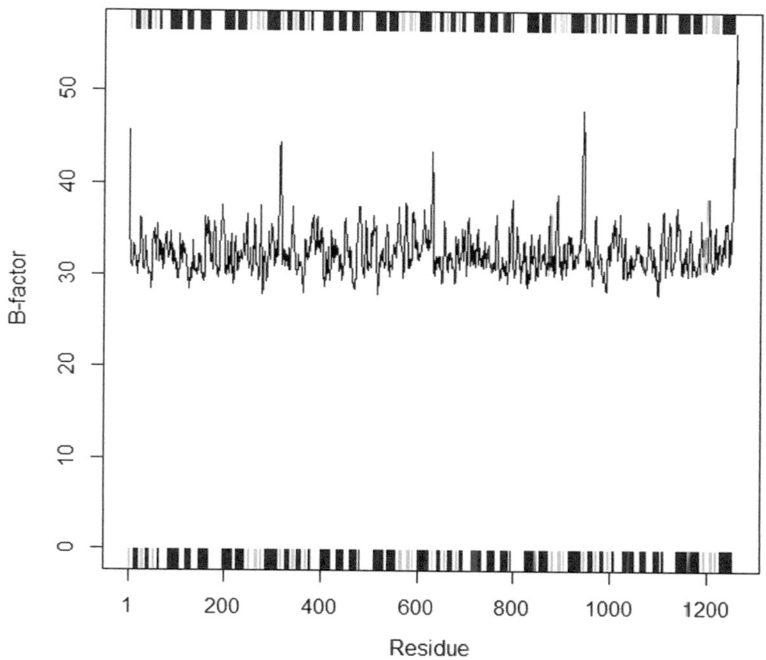

Fig. 5.7 Residue B-factors for malate dehydrogenase (PDB ID 2DFD) with secondary structure annotations in the marginal regions

2. Malate dehydrogenase and lactate dehydrogenase are homologous metabolic enzymes. Download the PDB structures of human malate dehydrogenase (ID: 2DFD) and human lactate dehydrogenase (ID: 5ZJE) in the R environment. Perform following tasks in the R environment using Package "bio3d":
   (a) Split and align the chains present in both dehydrogenase structures and compute the sequence identity and root mean square deviation (RMSD) between different chains of the protein structure.
   (b) Plot the histogram of RMSD and generate a dendrogram of RMSD clusters.

Solution (Figs. 5.8, 5.9, 5.10, and 5.11)

```
> library(bio3d)
> ids <- c("2dfd","5zje")
> raw.files <- get.pdb(ids)
trying URL 'https://files.rcsb.org/download/2dfd.pdb'
Content type 'application/octet-stream' length unknown
downloaded 920 KB
trying URL 'https://files.rcsb.org/download/5zje.pdb'
Content type 'application/octet-stream' length unknown
```

```
downloaded 2.5 MB
> files <- pdbsplit(raw.files, ids)
 > pdbs <- pdbaln(files,exefile="msa")
> pdbs$id <- substr(basename(pdbs$id),1,6)
> seqidentity(pdbs)
```

	2dfd_A	2dfd_B	2dfd_C	2dfd_D	5zje_A	5zje_B	5zje_C	5zje_D	5zje_E	5zje_F	5zje_G	5zje_H	5zje_I	5zje_J	5zje_K	5zje_L
2dfd_A	1	1	1	1	0.254	0.254	0.254	0.254	0.254	0.254	0.254	0.254	0.254	0.254	0.254	0.254
2dfd_B	1	1	1	1	0.254	0.254	0.254	0.254	0.254	0.254	0.254	0.254	0.254	0.254	0.254	0.254
2dfd_C	1	1	1	1	0.254	0.254	0.254	0.254	0.254	0.254	0.254	0.254	0.254	0.254	0.254	0.254
2dfd_D	1	1	1	1	0.254	0.254	0.254	0.254	0.254	0.254	0.254	0.254	0.254	0.254	0.254	0.254
5zje_A	0.254	0.254	0.254	0.254	1	1	1	1	1	1	1	1	1	1	1	1
5zje_B	0.254	0.254	0.254	0.254	1	1	1	1	1	1	1	1	1	1	1	1
5zje_C	0.254	0.254	0.254	0.254	1	1	1	1	1	1	1	1	1	1	1	1
5zje_D	0.254	0.254	0.254	0.254	1	1	1	1	1	1	1	1	1	1	1	1
5zje_E	0.254	0.254	0.254	0.254	1	1	1	1	1	1	1	1	1	1	1	1
5zje_F	0.254	0.254	0.254	0.254	1	1	1	1	1	1	1	1	1	1	1	1
5zje_G	0.254	0.254	0.254	0.254	1	1	1	1	1	1	1	1	1	1	1	1
5zje_H	0.254	0.254	0.254	0.254	1	1	1	1	1	1	1	1	1	1	1	1
5zje_I	0.254	0.254	0.254	0.254	1	1	1	1	1	1	1	1	1	1	1	1
5zje_J	0.254	0.254	0.254	0.254	1	1	1	1	1	1	1	1	1	1	1	1
5zje_K	0.254	0.254	0.254	0.254	1	1	1	1	1	1	1	1	1	1	1	1
5zje_L	0.254	0.254	0.254	0.254	1	1	1	1	1	1	1	1	1	1	1	1

Fig. 5.8 Sequence identity between different chains of human malate dehydrogenase and lactate dehydrogenase

```
> rmsd(pdbs, fit=TRUE)
```

	2dfd_A	2dfd_B	2dfd_C	2dfd_D	5zje_A	5zje_B	5zje_C	5zje_D	5zje_E	5zje_F	5zje_G	5zje_H	5zje_I	5zje_J	5zje_K	5zje_L
2dfd_A	0	0.176	0.252	0.209	3.682	3.771	3.801	3.763	3.671	3.692	3.68	3.689	3.837	3.781	3.688	3.754
2dfd_B	0.176	0	0.23	0.157	3.687	3.775	3.804	3.768	3.677	3.696	3.686	3.692	3.842	3.785	3.692	3.759
2dfd_C	0.252	0.23	0	0.211	3.674	3.758	3.786	3.748	3.663	3.682	3.67	3.678	3.824	3.766	3.679	3.742
2dfd_D	0.209	0.157	0.211	0	3.672	3.758	3.788	3.752	3.662	3.681	3.67	3.676	3.826	3.769	3.677	3.742
5zje_A	3.682	3.687	3.674	3.672	0	0.688	0.774	0.702	0.424	0.512	0.438	0.478	0.842	0.732	0.292	0.718
5zje_B	3.771	3.775	3.758	3.758	0.688	0	0.537	0.594	0.784	0.62	0.787	0.603	0.605	0.579	0.688	0.512
5zje_C	3.801	3.804	3.786	3.788	0.774	0.537	0	0.615	0.875	0.645	0.864	0.667	0.452	0.588	0.781	0.69
5zje_D	3.763	3.768	3.748	3.752	0.702	0.594	0.615	0	0.816	0.598	0.827	0.599	0.6	0.385	0.715	0.543
5zje_E	3.671	3.677	3.663	3.662	0.424	0.784	0.875	0.816	0	0.521	0.314	0.483	0.953	0.837	0.452	0.81
5zje_F	3.692	3.696	3.682	3.681	0.512	0.62	0.645	0.598	0.521	0	0.542	0.327	0.701	0.589	0.514	0.671
5zje_G	3.68	3.686	3.67	3.67	0.438	0.787	0.864	0.827	0.314	0.542	0	0.509	0.934	0.842	0.483	0.831
5zje_H	3.689	3.692	3.678	3.676	0.478	0.603	0.667	0.599	0.483	0.327	0.509	0	0.735	0.602	0.477	0.647
5zje_I	3.837	3.842	3.824	3.826	0.842	0.605	0.452	0.6	0.953	0.701	0.934	0.735	0	0.592	0.858	0.71
5zje_J	3.781	3.785	3.766	3.769	0.732	0.579	0.588	0.385	0.837	0.589	0.842	0.602	0.592	0	0.744	0.599
5zje_K	3.688	3.692	3.679	3.677	0.292	0.688	0.781	0.715	0.452	0.514	0.483	0.477	0.858	0.744	0	0.73
5zje_L	3.754	3.759	3.742	3.742	0.718	0.512	0.69	0.543	0.81	0.671	0.831	0.647	0.71	0.599	0.73	0

Fig. 5.9 Root mean square deviation (RMSD) between different chains of human malate dehydrogenase and lactate dehydrogenase

```
 > core <- core.find(pdbs)
core.inds <- print(core, vol=1.0)
> xyz <- pdbfit( pdbs, core.inds )
> rd <- rmsd(xyz)
> hist(rd, breaks=40, xlab="RMSD (Å)", main="Histogram of RMSD")
```

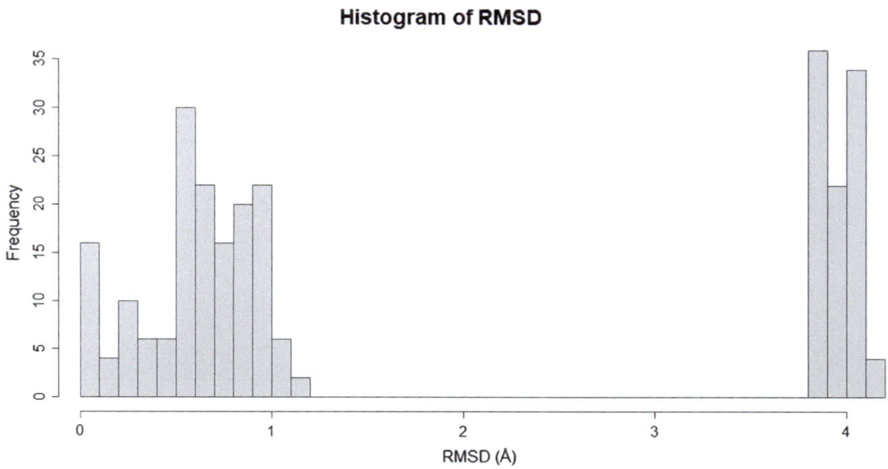

Fig. 5.10 A histogram of RMSD between different chains of human malate dehydrogenase and lactate dehydrogenase

```
>hc <- hclust(as.dist(rd))
> hclustplot(hc, k=3, ylab="RMSD (Å)", main="RMSD Cluster Dendrogram")
```

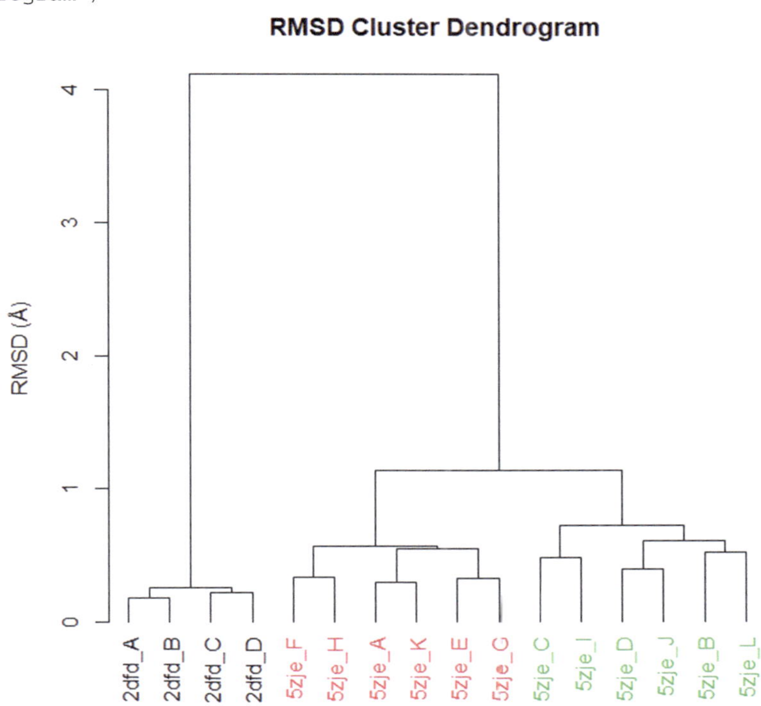

Fig. 5.11 A RMSD cluster dendrogram based on RMSD between different chains of malate dehydrogenase and lactate dehydrogenase

5.12 Multiple Choice Questions

1. Which of the following is NOT a molecular visualization application?
   (a) RasMol
   (b) PyMol
   (c) Modeller
   (d) jmol
2. Which of the following structure prediction method does NOT rely on a template structure?
   (a) Homology modelling
   (b) Ab initio modelling
   (c) Threading
   (d) Integrative modelling
3. The minimum percent identity required between a template structure and a target sequence for accurate homology modelling is:
   (a) 20%
   (b) 30%
   (c) 40%
   (d) 50%
4. Which of the following structure prediction software is based on threading?
   (a) I-TASSER
   (b) Modeller
   (c) SwissModel
   (d) HHPred
5. Which of the following interaction is modelled using Lennard-Jones potential during MD simulation?
   (a) van der Waals interaction
   (b) Electrostatic interaction
   (c) Both van der Waals and electrostatic interactions
   (d) None of the above
6. The potential energy function equation consists of following force(s):
   (a) Dihedral angle potential
   (b) Electrostatic interaction
   (c) van der Waals interaction
   (d) All the above
7. The global scoring function used by a docking program consists of:
   (a) van der Waals interactions
   (b) Hydrogen bonding
   (c) Electrostatic interactions
   (d) All the above
8. Which of the following statement is true for QSAR?
   (a) QSAR is useful in lead identification and optimization
   (b) QSAR predicts the ADMET properties
   (c) QSAR is based on linear or multiple regression method
   (d) All the above

9. A popular program for 6D-QSAR is:
   (a) Quasar
   (b) HQSAR
   (c) CoMFA
   (d) CoMSA
10. Which of the following is NOT a docking program?
   (a) Autodock
   (b) Glide
   (c) Bio3d
   (d) GOLD

Answer: 1. c 2. b 3. c 4. a 5. a 6. d 7. d 8. d 9. a 10. c

Box 5.1 B-Factors of a Protein Structure

The B-factors of a protein structure are also known as temperature structures and represent the flexibility of a protein due to fluctuation of C alpha atoms around their average positions. Due to this flexibility, there is a constant movement of polypeptide backbones and side chains of a protein molecule. The high B-factors indicate that the residue positions in the protein have a higher flexibility than the average value, whereas low B-factors are the reflections of rigid positions in the protein molecule. Buried residues in the core of the protein molecule are likely to have low B-values and therefore are more rigid in comparison to the residues present on the surface of protein.

Summary

- Homology modelling of a protein structure is performed using evolutionary related target protein structure from PDB.
- Ab initio modelling depends on energy functions or information obtained from other protein structures.
- Threading assigns folds in the target protein based common folds in the template PDB structures.
- Integrative modelling is an iterative process starting with a proposed initial model followed by several rounds of refinement and validation.
- Large-scale determination of protein structures for a systems level view of protein structural world is known as structural genomics.
- The CATH and SCOP databases describe protein structural classification covering the entire protein structures deposited in the PDB.
- Molecular dynamics simulations provide useful insights into mechanistic basis of protein folding, protein ligand interactions and protein dynamics.
- Molecular docking is useful in identification and optimization of lead compounds during a drug discovery process.

- Quantitative structure-activity relationships (QSAR) determine quantitatively the binding affinity of a drug-macromolecular target complex using linear or multiple regression methods.

Suggested Reading

Schwede T, Peitsch M (2008) Computational structural biology: methods and applications. World Scientific, Hackensack

Buxbaum E (2007) Fundamentals of protein structure and function. Springer, Cham

Becker OM, Karplus M (2006) A guide to biomolecular simulations. Springer, Dordrecht

Rigden DJ (2009) From protein structure to function with bioinformatics. Springer, Dordrecht

Lesk AM (2001) Introduction to protein architecture: the structural biology of proteins. Oxford University Press, Oxford

Molecular Evolution

6

Learning Objectives

You will be able to understand the following after reading this chapter:

- A general understanding of molecular evolutionary processes.
- Neutral and nearly neutral theories of evolution.
- The concepts of homology, molecular clock and genetic distances.
- A basic understanding of nucleotide substitution models.
- Phylogenetic tree and various methods of its reconstruction.
- Estimation of molecular signatures of natural selection.
- Evolutionary understanding of protein structures.
- Evolution of pathogenic viruses including human SARS-CoV-2.

6.1 Introduction

Evolution is a continuous process in natural populations in terms of regular fluctuations in gene (allele) frequencies. Some alleles show an increasing trend in frequency and reach up to 100% in order to be fixed in a population. On the other hand, other alleles tend to decrease in frequency and ultimately lost in a population. The molecular evolution is the result of accumulation of mutations in genes in form of substitution, insertion/deletion, recombination and gene conversion. Mutations generate many genetic variants in a population and this phenomenon is known as genetic polymorphism. The genetic polymorphism of a population is not maintained for indefinite period of time and new mutations are subjected to filtration by two evolutionary forces, namely natural selection and genetic drift, and ultimately fixed or eliminated in a population. The new morphological and physiological characters in a new species appear due to fixed mutations in its genome. The fixed mutations in

B. K. Tiwary, *Bioinformatics and Computational Biology*,
https://doi.org/10.1007/978-981-16-4241-8_6

a species are known as substitutions. The substitution rate is computed in terms of accumulation of genetic differences between individuals in an evolutionary time-scale. Consequently, substitution is the manifestation of differences in nucleotide or amino acid sequences between two taxa. Overall, the genetic divergence between two populations are generated by various factors such as underlying mutation rate, generation time and multiple evolutionary forces. When a particular allele is more fit than other alleles in an environment, it is subjected to positive selection; lower fitness of alleles exposes them to negative selection. In certain traits like sickle cell trait, heterozygote is more advantageous than the homozygotes in terms of protection against malaria. A genetic polymorphism is maintained in the population under the influence of balancing selection. In addition, rate of fixation of a mutation depends on the effective population size (Ne) producing offspring which is substantially smaller than the overall population size (N). The overall evolutionary process is a blend of both deterministic and stochastic processes. Natural selection operates on the gene frequency through differential reproduction and survival of some variants in an environment. Thus, evolutionary trajectory of selection can be predicted if prevailing environmental condition and fitness of the variant is known. In contrast, random genetic drift is a result of statistical fluctuations in gene frequency in a population with a small effective population size and ultimately determines the mutation rate. The change in gene frequency does not have a particular direction for genetic drift unlike *modus operandi* of natural selection. Nonsynonymous mutations are likely to be subjected to selective pressure and cause phenotypic changes in an organism. In contrast, synonymous mutations are largely neutral and are fixed in a population under the impact of genetic drift. The comparison of nonsynonymous and synonymous substitution rate gives us an assessment of positive and negative selection operating on genes.

6.2 Neutral and Nearly Neutral Theories of Evolution

Motoo Kimura proposed his landmark neutral theory of molecular evolution in 1968 which is central to the principles of molecular evolution. The rate of molecular evolution among proteins varies widely from the slowest evolving histone IV (10^{-11} substitutions per site per year) to the fastest evolving fibrinopeptide (9×10^{-9} substitutions per site per year). Based on this observation, Kimura and Ohta proposed that functionally less important molecules or parts of such molecules are likely to evolve faster than more important proteins. In other words, functionally important molecules or their parts are selectively constrained due to their detrimental effect on the fitness of the organism. The major evolutionary force involved in genetic substitutions is stochastic fixation of mutations. The majority of these substitutions are product of random fixation of neutral and nearly neutral mutations. The synonymous nucleotide changes and changes in nucleotides in introns, pseudogenes, other non-coding and nonregulatory regions are examples of neutral mutations. Some nonsynonymous mutations having no effect on the protein functions are also treated as neutral mutations. Since effective population size is usually very small in comparison to the magnitude of selection, positive selection plays an insignificant role in shaping the genome. Consequently, a minority of mutations are fixed under the influence of positive selection. Organisms are well adapted in their environment and many nonsynonymous mutations are deleterious to the organism. As a result, deleterious nonsynonymous mutations are readily removed by the purifying selection. Thus, Kimura advocated that evolution is largely a stochastic process operating on neutral mutations where substitutions are fixed by the random genetic drift. Tomoko Ohta proposed some modifications in the neutral theory as the nearly neutral theory of molecular evolution in 1973 and subsequently published two seminal papers in 1990 and 1991. Ohta emphasized on those borderline mutations which are neither strictly neutral nor are strongly selected and called them as nearly neutral mutations. In contrast, Kimura focussed only on the strictly neutral mutations. A mutation with selection coefficient (s) value much lower than the reciprocal of the population size behaves as a neutral mutation ($s = 0$) in a population. She further proposed that only neutral mutations are fixed in the population when s is less than or close to the reciprocal of the population size ($s << 1/2Ne$). On the other hand, natural selection rejects the mutations in the functionally constrained regions due to their detrimental effect on the fitness ($s >> 1/N$). There is a negative correlation between generation time and effective population size. Species with short generation time are likely to have large effective population size. The mutations vary with generation time and as a result, the differences in generation time and mutation rates cancel each other in order to make mutation rate effectively neutral. Thus, rate of molecular evolution appears to be proportional to absolute time rather than the generation time.

6.3 The Concept of Homology

The term homology refers to evolutionary relationship between traits in various organisms. Different kinds of traits such as structure and function of molecular sequences, genomes, chromosomes, cells, organs and behaviour may be evaluated for homology. The origin of all living beings on this earth can be traced back to a single common ancestor. Genes are treated as homologous (i.e. descent from common ancestor) due to their sequence similarity. However, the overall similarity between two genes gradually disappear as nucleotide or amino acid variation accumulates in both genes with passage of time. Thus, adequate similarity to declare them as homologous genes is retained in the sequence if they share a most recent common ancestor. Genes are either completely homologous or completely non-homologous. Thus, a statement that there is 70% homology between two genes/proteins is incorrect; instead, a statement that there is 70% similarity between two genes/proteins is correct. The higher similarity between two genes is an indicator of their likelihood of being homologous. There are different forms of homology based on evolutionary processes involved in the origin of a gene (Table 6.1; Fig. 6.1). Homologous genes in each species evolve independently due to a speciation event (cladogenesis) and are known as orthologous genes. For example,

Table 6.1 Different forms of homology and evolutionary processes involved

Homology	Evolutionary processes involved
Orthology	Speciation (Cladogenesis)
Pro-orthology	Speciation followed by gene duplication
Paralogy	Gene duplication
Inparalogy	Speciation and lineage-specific duplication
Outparalogy	Gene duplication and speciation
Xenology	Lateral gene transfer
Partial homology	Exon shuffling or other recombination events leading to a composite gene

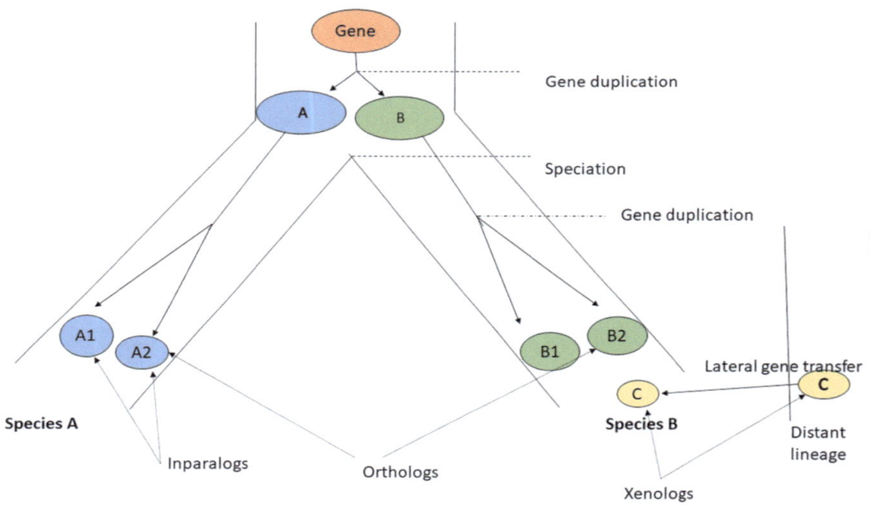

Fig. 6.1 Different forms of homology between genes based on speciation and duplication events

cytochrome C is an ortholog between human and chimpanzee. Moreover, the homology between a singleton gene and its duplicate genes in another lineage is known as pro-orthology. For instance, a homeotic gene AmphiOtx is pro-orthologous to Otx1 and Otx2 in bony fishes and tetrapods. In reverse, Otx1 and Otx2 are semi-orthologous to AmphiOtx. In contrast, genes originating from duplication event in a species are known as paralogous genes. Haemoglobin and myoglobin genes are well-known examples of paralogous genes in a vertebrate species. When gene duplication occurs after a speciation event, the paralogs are known as inparalogs or symparalogs. In contrast, gene duplication before the speciation event gives rise to outparalogs or alloparalogs. In some cases, homology appears due to lateral gene transfer in bacteria and is known as xenology. The homology between two genes may not be for the entire length of the gene but restricted to a specific part of the gene. This partial homology may be due to exon shuffling and other recombination processes resulting in generation of a composite gene. Fibronectin and tissue plasminogen activator genes in human show partial homology due to exon shuffling. Thus, orthologous and paralogous genes should be carefully selected for phylogenetic analysis in order to understand speciation and duplication events, respectively. Two non-homologous enzymes may secondarily acquire similarity at the active site during evolution due to similar functional constraint which is often mistaken for homology. This evolutionary process is known as convergent or parallel evolution.

6.4 Genetic Distances

The genetic or evolutionary distances between all pairs of aligned sequences are computed in form of a distance matrix. A pair of sequences derived from a common ancestor are likely to diverge in the course of time under the influence of various evolutionary forces. This divergence between two sequences is measured by genetic distance. Genetic distance is an indicator of similarity between sequences and provides useful information for inference of a phylogenetic tree. The substitution process is generally modelled as a stochastic or random event. One needs to specify the appropriate statistical model of substitution prior to computing genetic distances between two nucleotide or protein sequences. The simplest measure of genetic distance (d) is the total number of nucleotide differences per site between two sequences. This measure is also known as the p-distance or the observed distance. However, it is not informative about the actual number of substitutions occurred in case of high level of divergence between two sequences. Multiple nucleotide substitutions at a particular site in an evolutionary timescale usually leads to the saturation of such sites in the DNA sequences.

Another approach to measure expected genetic distances between a pair of sequences is to apply a likelihood function L (d). This likelihood is defined as the probability to observe two sequences given the distance d. The value of the distance d that maximizes L (d) is known as the maximum likelihood estimate (MLE) of the genetic distance.

6.5 Nucleotide Substitution Models

There is a variety of nucleotide substitution rate matrices from a very simple model like Jukes–Cantor (JC69) model to many highly complex models. In JC69 model, the equilibrium frequencies of four nucleotides are equal (25%) and each nucleotide has the same probability to replace other one. In the general time reversible (GTR) model, all eight free parameters of a reversible nucleotide rate matrix are clearly stated. The free parameters are usually unknown and required to be estimated from the data. Thus, it is always desirable to reduce the number of free parameters by including some constraints in the underlying substitution process. The Tamura–Nei (TN93) model considers only two independent rate parameters, the ratio (κ) between transition and transversion rates and the ratio (γ) between two types of transition rates. If we assume that the purine and pyrimidine rates are equal ($\gamma = 1$), The

Tamura–Nei model becomes a simpler model and the reduced model is known as HKY85 model. This HKY85 model can be further reduced to the Kimura two-parameter (K80) model assuming uniform nucleotide frequencies of 25% for each nucleotide base. The HKY85 model is reduced to F81 model and K80 model is reduced to the simplest model, Jukes and Cantor model in case the $\kappa = 1$. It is always advisable to choose a simple model like Jukes–Cantor model or Kimura two-parameter model than a complex model like GTR model for phylogenetic analysis of closely related nucleotide sequences. The best-fit evolutionary model of a particular dataset is selected using different statistical approaches such as hierarchical likelihood ratio test, Akaike Information Criterion (AIC) and Bayesian Information Criterion (BIC). The likelihood ratio test (LRT) compares the log likelihoods of two contrasting nested models to find the best fit and is defined by $LRT = 2\,(l_1 - l_0)$, where l_1 and l_0 are the maximum likelihood values under complex (parameter-rich) and simple models (less parameter rich), respectively. However, AIC and BIC are applicable to both nested and non-nested models. The fully automated model selection process to find best-fit model is implemented for nucleotide sequence and protein sequence in programs jModelTest and protTest, respectively. The protTest selects the best model of amino acid replacement. These evolutionary models are based on the amino acid substitution matrices computed from a large dataset of protein families. The common amino acid replacement models used for protein sequence analysis are Jones–Taylor–Thornton (JTT), Dayhoff, BLOCKS Substitution Matrices (BLOSUM62) and Whelan and Goldman (WAG) matrices.

It is well known that nucleotide substitution rates vary greatly at different positions in DNA sequences. For instance, the third codon position mutates faster than the first and second codon positions. This rate heterogeneity at different codon positions may influence the measurement of genetic distance. A gamma (Γ) distribution model with variance $1/\alpha$ considers varying degree of distribution of rates over sites by adjusting the shape parameter α. The gamma distribution is bell shaped in case the value of α is more than one and shows weak heterogeneity over sites. On the other hand, the gamma distribution looks like L-shape if the value of α is less than one and represents strong rate heterogeneity over the sites. Four to eight discrete categories of gamma distribution are usually considered in phylogenetic analysis to approximate the continuous gamma distribution.

6.6 Phylogenetic Analysis

The evolutionary relationship between organisms and their genes/proteins are represented in form of a phylogenetic tree. A phylogenetic tree resembles the parts of a tree such as root, branch, node and leaf (Fig. 6.2). The existing or extant taxa are represented by the terminal leaves or nodes and commonly known as the operational taxonomic unit (OTU). The internal nodes are known as hypothetical taxonomic units (HTU) because these nodes are the hypothetical ancestors of OTU. A phylogenetic tree consists of two characteristic features, namely topology of a tree and its

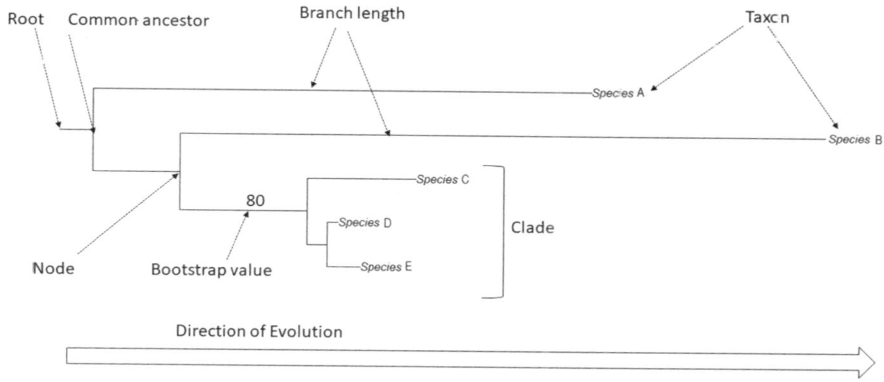

Fig. 6.2 A rooted phylogenetic tree with its constituent parts

branch lengths. Topology of a phylogenetic tree is represented by its branching pattern. There may be the presence of root (i.e. rooted) or absence of root (i.e. unrooted) in a phylogenetic tree. An unrooted phylogenetic tree only provides the details regarding the topology and the branch lengths. However, it lacks very vital information about the evolutionary history of sequences under study. An unrooted phylogenetic tree is usually transformed to a rooted phylogenetic adding additional evolutionary information such as a reference sequence or an outgroup, that is, the most distant species than the rest of the species (ingroup). A phylogenetic tree is, in general, bifurcating in nature. But, multifurcating phylogenetic trees are also reconstructed because of some statistical errors or presence of a zero-branch length. A species tree describes the phylogenetic history of a species, whereas the gene tree is generated from a particular gene sequence and represents the evolutionary history of that particular gene. Thus, the overall topology of an individual gene tree is likely to differ from a species tree. The branch lengths of a phylogenetic tree are represented as the number of substitutions per site. In a phylogram, the branch lengths are scaled based on the evolutionary distances between taxa. On the other hand, a chronogram demonstrates the complete evolutionary timescale from the common ancestor to the extant species.

6.7 Methods for Reconstruction of Phylogenetic Tree

Various statistical methods are available to reconstruct a phylogenetic tree from nucleotide and protein sequence data (Tables 6.2 and 6.3). Most popular methods may be classified into four major categories: distance-based methods, maximum parsimony methods, maximum likelihood methods and Bayesian inference methods.

Table 6.2 Software for common evolutionary analyses

Method	Name of the program	URL
Phylogenetic analysis	MEGA	www.megasoftware.net
	PHYLIP	evolution.genetics.washington.edu/phylip.html
	DAMBE	http://dambe.bio.uottawa.ca/DAMBE/dambe.aspx
	phyML	http://www.atgc-montpellier.fr/phyml
	RaxML	cme.h-its.org/exelixis/web/software/raxml
	MrBayes	nbisweden.github.io/MrBayes/download.html
Recombination detection	RDP4	http://web.cbio.uct.ac.za/?darren/rdp.html
	GARD	http://www.datamonkey.org/GARD
Positive selection	PAML	http://abacus.gene.ucl.ac.uk/software/paml.html
	HyPhy (BUSTED, aBSREL, FADE)	www.hyphy.org
Relaxed selection	HyPhy (RELAX)	www.hyphy.org
Substitution saturation	DAMBE	http://dambe.bio.uottawa.ca/DAMBE/dambe.aspx
Synonymous constraint	FRESCo	A batch script for HyPhy
McDonald–Kreitman test	iMKT	https://imkt.uab.cat
Positive selection in populations	DnaSP	http://www.ub.edu/dnasp

Table 6.3 Popular R packages for phylogenetic analysis

Package	Phylogenetic application
Ape	A core package for phylogenetic analysis and evolution
Phangorn	A package suitable for maximum likelihood analysis
Treedist	A package to measure similarity between phylogenetic trees
Ggtree	A package for visualization and annotation of phylogenetic trees
Treeman	An R package for manipulation of phylogenetic trees
Phytools	A package for phylogenetic comparative biology

6.7.1 Distance-Based Methods

The distance-based methods make an attempt to fit a phylogenetic tree to a matrix of pairwise genetic distances computed from sequence data. Here, the number of substitutions occurred truly in the sequences are estimated using a specific evolutionary model best fitting the data. Distance-based methods are implemented in two steps. First, the pairwise evolutionary distances are computed and stored in a matrix of pairwise distances. The second step involves clustering algorithms to infer the tree topology based on evolutionary distances. Clustering methods such as UPGMA (unweighted-pair group method with arithmetic means) usually generate an ultrametric tree assuming molecular clock. The ultrametric trees are the rooted trees with all the end nodes having equal distance from the root. The tree is constructed in sequential order by clustering those OTUs closely related to each other followed by more distant OTUs. However, the UPGMA method is highly sensitive to errors due to unequal rates of substitutions in different lineages. Since the ordinary clustering methods have serious limitations, additive distance trees are the better alternatives to ultrametric trees. Additive distance trees are unrooted trees in which the genetic distance between a pair of OTUs is equal to sum of the lengths of connecting branches. This kind of tree is always a better fit to the genetic distances in the absence of clock-like behaviour in the sequence data. An additive tree can be constructed using minimum evolution (ME) method that minimizes the length of the tree which is the sum of the lengths of its branches. The length of the tree is computed from a matrix of genetic distances. The main drawback of this method is to search all possible topologies to find the minimum tree which appears to be an impractical computational task with more than ten sequences. A better solution to this computational challenge is to use a heuristic approach to find the ME tree using clustering without assuming a clock-like behaviour. This improved method is known as the neighbour joining (NJ) which is most popular distance-based method used in phylogenetics. The distance-based method can be best adopted for phylogenetics combining the NJ approach with bootstrapping. The reliability of an inferred tree especially a specific clade of a tree is usually evaluated using bootstrap analysis. It is very efficient statistical method to approximate underlying sampling distribution by resampling from the original data. In phylogenetics, bootstrapping is implemented on the original alignment by sampling of nucleotides sites with replacement, reconstructing the phylogenetic tree and checking the presence of the original nodes in all resampled set of phylogenetic trees (Fig. 6.3). This process is usually repeated from 200 to 2000 times and the proportion of each node among all bootstrap replicates is known as the bootstrap value. Bootstrap value indicates the statistical confidence supporting a monophyletic clade and usually labelled on the top of each clade. A bootstrap value of 70 or more demonstrates a good confidence for a specific clade and the clades having lesser bootstrap values are treated with caution and usually collapsed into a multifurcated phylogenetic tree. Jackknifing is an alternative approach to bootstrapping in evaluating specific clades in a phylogenetic tree. Here, half of the sites are randomly purged from the original sequences to generate numerous new samples. The proportion of clades in the resampled tree is known as the jackknife value and 70% or more jackknife value of a clade is treated with good confidence. The NJ method is very fast even for phylogenetic analysis of hundreds of sequences and is an appropriate method for reconstructing large trees.

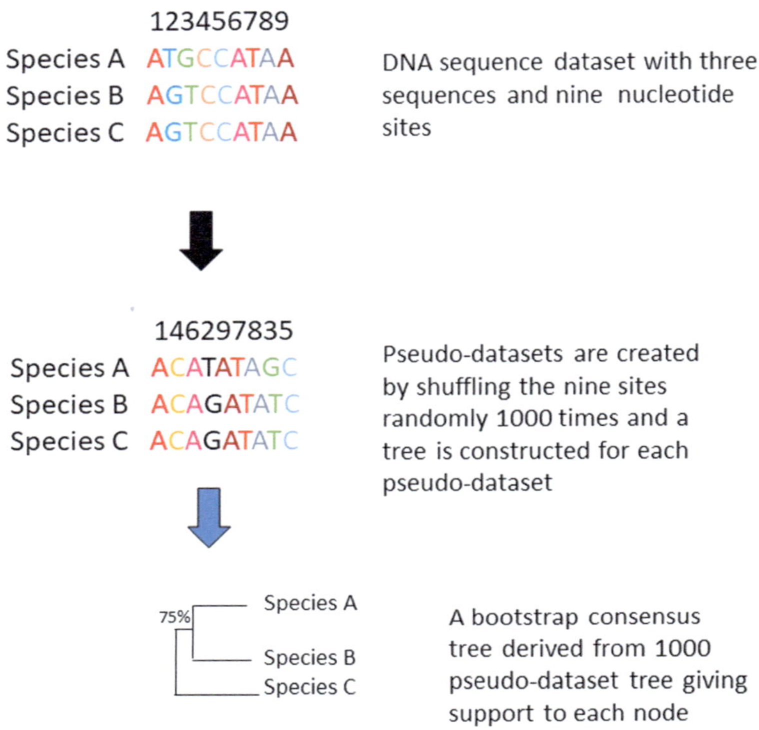

Fig. 6.3 The steps involved in bootstrapping

6.7.2 Maximum Parsimony

It is one of the best and useful method for molecular phylogenetics based on discrete character states and is not based on an explicit model of evolution. This method searches for tree or a collection of a tree assuming minimum number of genetic changes from a common ancestor to its descendants. This method is equally effective in case the rate of evolution is either fast or slow. This method changes randomly the topology of a tree in stepwise manner until a parsimony score is obtained which cannot be improved further. The exhaustive search for the best tree is done by evaluating all possible trees but this algorithm is feasible up to 11 sequences. An alternative search method known as the branch-and-bound method is effective for alignments containing 12 to 25 sequences. This method can be improved further adding a heuristic method or neighbour joining. Heuristics is an approximate method which can make an effort to find optimal solutions. First, an initial tree is generated using a greedy algorithm and then subjected to several rounds of perturbations to yield the most optimal tree. The stepwise addition is the most widely used greedy algorithm to generate an initial tree but rarely yields a globally optimum tree. Thus, three common methods of tree-rearrangement perturbations or branch swapping are nearest-neighbour interchange (NNI), subtree pruning and regrafting (SPR) and tree bisection and reconnection (TBR). These methods are effective for an alignment up to 100 sequences. However, a hill climbing algorithm like branch swapping is likely to be trapped in local minima.

The major advantage associated with this method is that it is free from unrealistic model assumptions. The limitation of this method is high computational cost, oversimplification of model of evolution and is likely to be inaccurate in case of substantial evolution. Parsimony method takes into account only the informative sites during phylogenetic analysis and does not consider information from the non-informative sites. Sometimes, maximum parsimony method is likely to generate an incorrect topology due to artificial clustering of long or short branches forming a common clade. This phenomenon is known as *long branch attraction* and short branch attraction, respectively.

6.7.3 Maximum Likelihood Methods

Likelihood is a statistical measure of goodness of fit of a specific model to some datasets under given values of unknown parameters. In phylogenetics, maximum likelihood finds for the best tree among a set of competing hypotheses such as possible tree topologies, branch lengths, parameters describing the model of sequence evolution, etc. The probability of the data is computed under assigned values of the parameter and a decision is made regarding their likelihood. It is expected that some hypotheses generate the observed data with a higher probability than other alternatives. When maximum likelihood estimate ($\widehat{\theta}$) is chosen, the observed data are produced with the highest likelihood. However, likelihood function differs from probability, although it is defined in terms of probability. Likelihood is the probability of the observed event instead of unknown parameters and is not linked to the probability that the specified model is correct.

Since points mutations are considered as random events, the probability of finding a point mutation along a lineage in a phylogenetic tree can be computed in maximum likelihood framework. Here, we determine the tree topology, branch lengths and parameters of the substitution model that maximizes the probability of observing the sequence data under study. In fact, the likelihood L is the conditional probability of a sequence data given a specific model of substitution with a set of parameters θ and the tree τ with branch lengths:

$$L(\theta, \tau) = \text{Probability (Sequence alignment} | \text{ tree, evolutionary model)}$$

It is computationally impossible to test each tree from numerous possible trees and estimate their model parameters. Therefore, different greedy heuristics using hill climbing are commonly used to search for the maximum likelihood of a tree. Stepwise addition is one of the early heuristics used and is available as DNAml in PHYLIP package. Other popular heuristics commonly used are full-tree rearrangement operations using three approaches: nearest-neighbour interchange (NNI), tree bisection and reconnection (TBR) and subtree pruning and regrafting (SPR). The certainty of a maximum likelihood tree can be assessed by comparing maximum likelihood values using likelihood ratio test (LRT). Similarly, statistical support to a specific branch is assessed by either nonparametric bootstrapping or jackknifing.

6.7.4 Bayesian Phylogenetic Inference

The Bayesian phylogenetic inference is based on our prior beliefs about the topology of a phylogenetic tree. In case the background information regarding the topology of a tree is lacking, equal probability is assigned to each possible tree topology and such prior probability is known as uninformative prior. We need a molecular sequence alignment data and an evolutionary model in order to update this prior probability to a posterior probability distribution. Bayes' rule is implemented to compute the posterior probability distribution. The posterior probability stipulates the probability of an individual tree given the prior, model and the data. The most of the posterior probability focusses on a single tree or few possible trees out of a large tree space in case the sequence data is informative. It is an impossible task to compute posterior probability analytically or drawing random sampling because posterior probability is restricted to a very small part of the huge parameter space. Therefore, the posterior probability distribution (Fig. 6.4) is usually estimated using the Markov Chain Monte Carlo (MCMC) sampling. Markov chains have a tendency to converge towards an equilibrium state irrespective of starting value. Thus, our focus is to set up a Markov chain converging on our posterior probability distribution. All plausible trees are sampled in order to find the range of a parameter, e.g. substitution rates in MCMC analysis. The MCMC algorithm samples the entire distribution avoiding any suboptimal solution. It proposes a new set of parameter values including tree topology in every step (generation) and the newly proposed state is either accepted in case better than previous state or rejected if worse than the previous state. The present state is retained in case of rejection and the process is repeated. The simulation process continues for a long duration changing one tree topology to another topology. The early phase of the simulation run is known as ***burn in*** and samples of this period is discarded to avoid the influence of starting value. The posterior clade probabilities are used as a statistical support to a clade in Bayesian phylogenetic inference in contrast to the bootstrap value used in other phylogenetic methods.

6.8 Molecular Clock

The molecular clock hypothesis is based on the assumption that the rate of molecular evolution is approximately uniform over evolutionary timescale. This hypothesis is fully consistent with the arguments of neutral theory of evolution. Emile Zuckerkandl and Linus Pauling proposed this concept in 1962 by comparing the amino acid changes in the various protein sequences including haemoglobin in

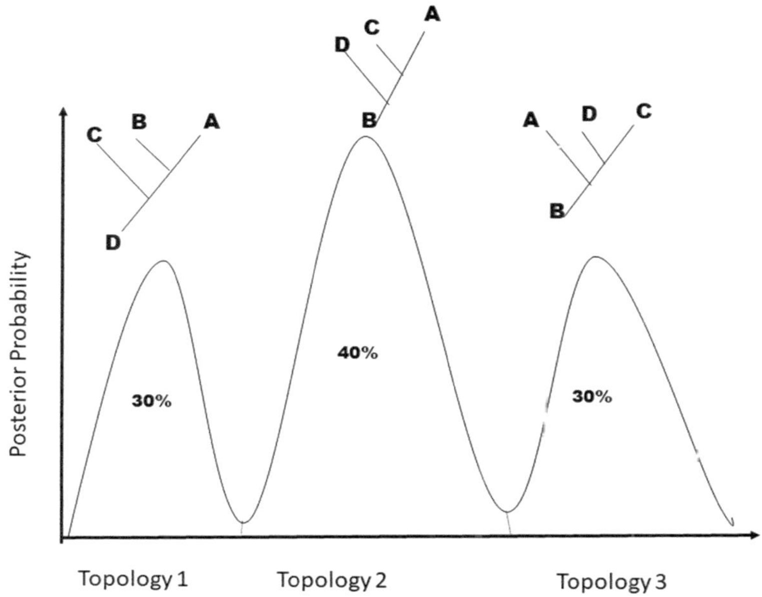

Fig. 6.4 A multimodal posterior probability distribution showing one-dimensional representation of three different phylogenetic tree topologies. The area under each topology is the posterior probability of respective topology

different species with their age estimated from fossil records. Overall, there was a linear trend in amino acid differences in respect to evolutionary time. Although there was a constant rate of molecular evolution across the species, each protein had a characteristic rate of molecular evolution (Fig. 6.5). The fibrinopeptides had a faster rate of molecular evolution in comparison to extremely slow rates in cytochrome c and histones. The differences in characteristic rates of evolution of these proteins can be explained by the proportion of neutral sites in the protein sequences. The proteins showing faster rate of evolution are likely to have a greater proportion of neutral sites. The molecular clock ticks at irregular intervals and is probabilistic in nature. The molecular clock estimates are prone to two types of errors. First type of error is caused by the sloppy nature of the rate of substitution process leading to a huge imprecision in the molecular date estimation. The variable mutation rate, population size and proportion of sites under differential selection pressure are the underlying causes of the second type of error. The presence of global molecular clock in a gene is usually tested by relative rate test, Tajima test and the likelihood ratio test. Since the global molecular clock is usually lacking in majority of genomic sequences, the presence of local molecular clock may be tested in specific lineages allowing rate variation across the lineages (relaxed clock). Molecular clock is useful in estimating the date of origin of a viral disease in absence of any fossil record. The evolutionary history of an epidemic disease is reconstructed from available extant virus samples. The viral molecular clocks are usually calibrated using the stored viral samples.

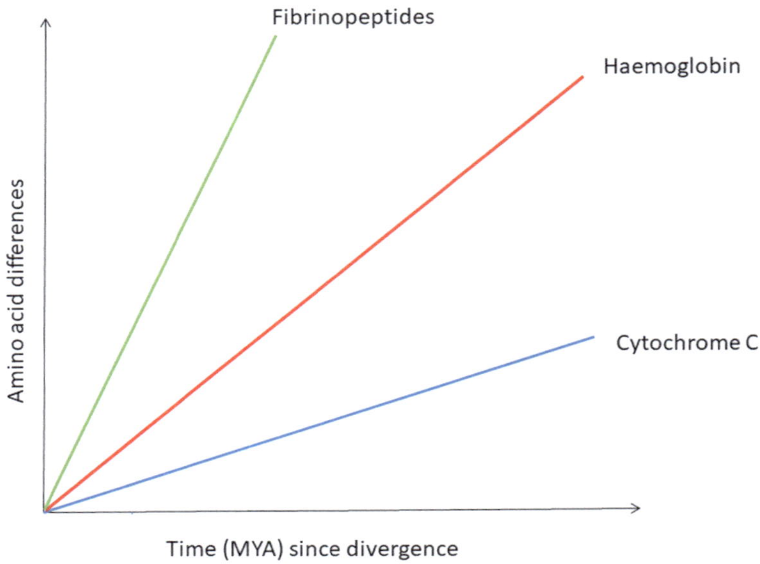

Fig. 6.5 The molecular clock of three different proteins ticking at variable rates

6.9 Ohno Hypothesis

The evolution of organismal complexity and the origin of novel gene function are likely to be linked with gene duplication and whole genome duplication. Susumo Ohno proposed his hypothesis that duplication of genes and whole genomes have played an important role in the evolution of organismal complexity. He further suggested that the duplication of a gene is followed by retention of original function by one copy and adoption of new function by another copy. The paralogs that emerged in the vertebrate genome through whole genome duplication are known as Ohnologs. The expansion in the number of genes in genomes is caused by various processes such as whole genome duplication, tandem gene duplication and segmental duplication. Consequently, large genomes allow more functional diversification of genes and creation of gene families leading to formation of complex gene networks in an organism. For example, amphioxus, ancestral cephalochordate lineage, usually have small number of gene families even singleton in many cases. On the other hand, advanced vertebrates such as mammals have three to four copies of genes per family. The two-rounds of genome duplication (2R) hypothesis was proposed as an explanatory model for the presence of large number of gene families in vertebrates. This model suggests that early genome duplication occurred in the genome of a cephalochordate-like ancestor (1R) followed by a second round of genome duplication (2R) leading to the formation of four sets of genomes. This 2R model is based on the empirical evidences obtained during genomic analysis of deuterostomes. Evolution of Hox clusters from one cluster in amphioxus to four clusters in tetrapods is a very strong evidence in support of this model. In addition, the synteny (i.e. same gene order) of gene clusters across evolutionary distant vertebrates also supports this hypothesis. Interestingly, an additional fish-specific whole genome duplication (3R) occurred after 2R duplication. For example, teleost fishes have multiple copies of some genes and gene clusters, whereas tetrapods have only one copy of the same in their genomes.

6.10 Molecular Signatures of Selection

Natural selection frequently acts on the genes and genomic sequences in an organism and thereby causes some genomic variations that are likely to enhance the overall fitness of individuals carrying these variations. These molecular footprints left on the nucleotide and protein sequences by the natural selection in the past can be identified using various statistical approaches. The relative rate of silent and replacement fixations in the molecular sequences is an important information to infer the past action of natural selection. A mutation in the gene, if confers a fitness benefit to the species is likely to be under positive selection that includes different processes. Positive selection is usually restricted to a small region of a gene and may point to a functionally important region or sites of a selected gene. There are two primary forms of positive selection: directional and diversifying selection. The action of directional selection on a particular gene is manifested in form of concerted substitution in the direction of a specific amino acid residue which is ultimately fixed in a population after a long span of time (selective sweep). The development of a specific mutation conferring drug resistance in HIV-1 against retroviral drugs is an example of selective sweep. On the other hand, diversifying selection maintains an amino acid diversity at a specific site in a population. This type of natural selection operates on the codon positions of HIV-1 that are the targets of the human immune response. There are two major approaches to identify natural selection on a gene. In order to perform selection analysis on a certain gene, we need a multiple sequence alignment of orthologs from different species or strains along with underlying phylogenetic tree of these species or strains. In case of recombination, each non-recombinant segment of the gene is represented by a separate phylogenetic tree. One needs to be very careful during multiple sequence alignment to preserve all codons and avoid any frameshift. There are two methods to identify positive selection on a gene based on the predictions of Kimura's neutral theory of evolution. The first approach deals with comparison of nonsynonymous and synonymous changes in a gene. The synonymous (silent) changes are likely to be neutral and a greater rate of nonsynonymous (amino acid replacement) changes relative to synonymous changes indicate the action of positive selection. It is estimated as ω or dN/dS or Ka/Ks ratio for the coding sequences. The ω is measured as the observed number of nonsynonymous substitutions per nonsynonymous site (dN) over the synonymous substitutions per synonymous site (dS) (Fig. 6.6). The ω is expected to be equal to one under the assumption of neutral evolution due to the fact that the synonymous changes are

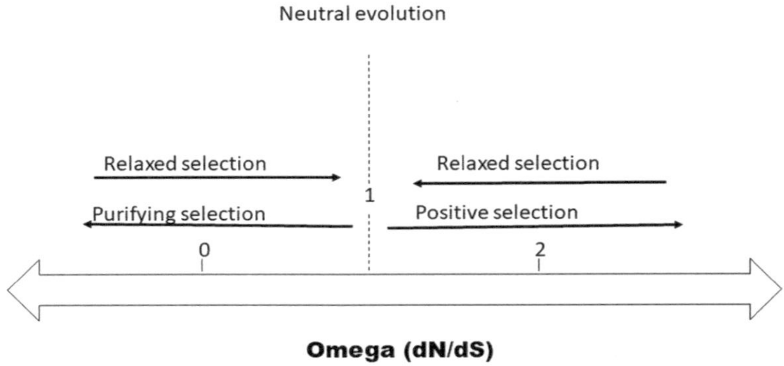

Fig. 6.6 The directional shift in ω values away from or towards the neutral value ($\omega = 1$) indicates the impact of natural selection on a gene. Positive selection operates in case of ω value more than one, whereas purifying selection shift the ω value towards zero. Relaxed selection is identified in a lineage by a shift of ω towards the neutral value in comparison to another related lineage

more or less equal to amino acid changes. The essential protein sequences allow only minor replacement changes in the amino acid sequences and majority of amino acid changes are eliminated by natural selection. This kind of natural selection is known as the negative or purifying selection and estimated as an ω value significantly less than one. An ω value close to zero indicates an exceptional selective constraint on the protein sequence. However, amino acid replacements for some proteins are also beneficial for the organism under selection. The genes encoding these proteins show an ω value significantly more than one and the selection operating is inferred as positive or directional selection. A high value of ω indicates the recurrent rounds of selection on a gene rather than a single selective sweep. The genes under positive selection are usually involved in male reproduction (e.g. human protamine gene), host's immune response to pathogen (e.g. human major histocompatibility complex (MHC) locus) and neural developments in primates. The estimation of ω varies from site to site and from branch to branch in a phylogenetic tree during a selective event. The positive selection is estimated based on the variation of ω across the sites and/or lineages. In site models, two maximum likelihood models, the first model (positive selection) allowing a class of codons in the alignment to evolve with ω more than one and the second model (neutral model) not allowing the same, are compared using likelihood ratio tests. In case, two models are found to be significantly different based on the chi-square distribution, the neutral model is rejected in favour of positive selection model. In branch-site models, episodic positive selection is detected on specific branches of a phylogenetic tree. Relaxed selection or relaxation of positive or purifying selection is also identified on a specific lineage as a shifting of ω value towards one in comparison to a closely related lineage.

When molecular sequences are sampled from different individuals in a population, the ω value reflects the ratio of replacement to silent polymorphism in a population. The ω value is likely to be one in the absence of positive and negative selection and any significant statistical deviations in this value indicates the action of natural selection. Natural selection might not have fixed the beneficial mutation or removed the deleterious mutation in a population during recent evolutionary history. Thus, the McDonald–Kreitman test, which compares the distribution of nonsynonymous and synonymous polymorphisms in a particular lineage with the ratio of nonsynonymous to synonymous differences fixed between lineages or species, is widely used to detect recent positive selection. However, saturation of substitution rates especially at the synonymous site for deep tree branches may affect the estimation of selection in case of fast evolving sequences. However, branch-site models are insensitive to the biases introduced by synonymous site saturation and are suitable for analysis of distant species. In addition, there are certain indexes to detect the saturation indexes and third codon position is often removed from analysis in case of synonymous site saturation. Moreover, this test is extremely conservative in the assessment of positive selection as it is an average estimate over the entire gene. This limitation can be overcome using the entire phylogeny instead of sequence pairs and additional computational cost. The purifying selection or positive selection are relaxed on certain genes by shifting the ω values towards the value of one (neutral value). The relaxed selection is determined based on whether intensity of natural selection is relaxed or intensified in two subsets of branches of a phylogenetic tree. The selection pressure either significantly increases or decreases in a foreground lineage with respect to a background lineage and is measured as a model parameter K (relaxation parameter).

The second approach is based on the predictions made by the neutral theory on allele or haplotype frequency in a particular population or between two populations. For example, if positive selection has swept a mutation to high frequency in a certain genomic region, it is expected that the targeted genomic region is likely to have low sequence diversity, a surplus of rare alleles and a greater amount of linkage disequilibrium than the expectation of neutral theory. Selection operating on a particular population in comparison to other populations eventually leads to a greater degree of population differentiation than expected in neutral evolution. Most common test statistics are Tajima's D, Wright's FST and Fu's W to measure these patterns in a population. However, demographic history, population bottleneck and expanding human population may generate spurious signatures of positive selection using this approach.

6.11 Evolution of Protein Structure

The evolution of a protein structure can be explained through a sequence space. Each possible amino acid sequence of a protein is treated as node and a single amino acid mutation between each sequence is represented by an edge in a sequence space. Since each node has a specific physical and functional feature, a genotype-phenotype space is created having all possible protein sequences. The protein sequence follows the edges in evolutionary trajectories through genotype-phenotype space. This approach helps us in understanding the underlying evolutionary mechanism across genotype-phenotype space in emergence of diversity and novel functions of proteins. These evolutionary trajectories of proteins can be traced using combined computational and experimental approaches.

The reconstruction of an ancestral state of an extant protein or a group of proteins is one of the powerful methods of choice. Firstly, the ancestral protein reconstruction was performed at the internal node of a phylogenetic tree using maximum likelihood approach. The maximum likelihood sequence at the internal node has a high probability of generating present world protein sequences at the terminal ends of a tree. Finally, the genes encoding the ancestral state of protein are synthesized and expressed in vitro cultured cells. Thus, the structural and functional analysis of each resurrected protein will allow us to understand the impact of historical mutations on the biophysical property of a protein structure.

Another powerful strategy is the directed evolution of a protein towards a desired function. Here, a library of random variants of a desired protein is generated followed by screening those variants with the specific property. The screened variants are subjected to repeated mutations to select the protein with the optimal function. The underlying mutations present in each variant of a protein and their impact on the overall evolution of a protein structure can be well understood using this approach.

The third strategy is focussed deeply on a part of the sequence space without any knowledge about the ancestral state. Here, random mutagenesis is performed on a protein followed by weak selection on the mutants for a desirable property. Consequently, the library of clones with desired property is enriched and the clones without this property are depleted. Finally, we can quantitate the direct and epistatic effects of each mutation based on the degree of enrichment of each clone after deep sequencing. Thus, this approach will not only unveil the distribution of a desirable sequence in the sequence space but also decipher the role of key evolutionary forces in directing trajectories across the sequence space.

Most protein structures are found to be marginally stable. The stability of a protein structure is described by the difference in free energy between its unfolded and folded states. A majority of proteins are marginally above the energetic threshold of unfolding and their stability can be improved further by simple point mutations. It is assumed that marginal stability of a protein structure is a trade-off between stability and function. However, it is possible to create an enzyme through directed evolution which is hyperstable as well as hyperactive refuting this trade-off hypothesis. Recent studies have shown that the marginal stability of a protein structure is a product of neutral evolution through balance between mutation and selection. In case, the hyperstability of a protein has neither beneficial nor deleterious effect on the function, the natural selection will not be able to discriminate between marginal and hyperstable ones. Thus, both mutation and random genetic drift are likely to drive the proteins neutrally towards lower stable states compatible with their function. Protein structures are maintained by some physicochemical constraints like correct folding, solubility, maintaining a particular function and thermodynamic stability.

Parallel evolution of phenotypes in different lineages occurs very frequently in nature under similar selective constraints. Repeated occurrence of similar mutations as a determinant of protein structure throws some light on the fundamental question in evolution: Is molecular evolution repeatable, predictable and deterministic in nature? Interestingly, parallel evolution occurs in independent proteins by accumulating same mutations possibly following same order. Rat and mice across different continents developed same mutations in the vitamin K epoxide reductase complex when exposed to warfarin. The large effect parallel mutations often occur at the key functional sites of the protein such as catalytic site of enzyme and binding site of a ligand. Moreover, these mutations are accompanied by non-parallel secondary mutations with small effect located far from the functional site. The non-parallel nature of secondary mutations indicates a weaker selective constraint on these mutations.

The emergence of unique physical and biological properties in a protein such as thermodynamic stability and substrate specificity is a result of interactions among amino acid residues. Single permissive mutation does not have any impact on the protein function per se but need one or more mutations for altered function in the protein. Thus, it allows a mutation to change the function of a protein by adopting contingency in the evolutionary process. A single mutation (H274Y) in the neuraminidase of influenza virus, although, confers resistance against the antiviral drug oseltamivir but adversely affects the viral fitness. Two permissive mutations shortly before the appearance of deleterious mutation counteract the adverse effect of this mutation. Permissive mutations also act as a buffer against the deleterious function-switching mutations. Usually, these new functional variants changing the protein function are removed frequently by the purifying selection. These permissive mutations are also observed in the evolution of new enzyme functions using directed evolution.

Permissive mutations operate using two physical mechanisms: non-specific epistasis and specific epistasis. The improvement of influenza virus fitness is an example of non-specific epistasis. The function-switching mutations are found to adversely affect the stability of a protein. But, permissive stabilizing mutations improve the stability of protein and shift the evolutionary trajectory towards new function. For example, the active site of beta lactamase is distended in bacteria to degrade cephalosporin antibiotics. Although, the expansion of active site increases the activity of enzyme but decreases its stability leading to weak resistance. Moreover, a high resistance variant of enzyme is developed by a compensatory mutation far away from active site. The second class of permissive mutations is based on specific epistasis where two and more mutations collaborate to produce new functional protein sometimes through conformational changes. For example, glucocorticoid receptor has undergone two historical mutations to develop specificity to glucocorticoid hormones in its evolutionary history. First mutation introduced a hydrogen bond on the helix in the ancestral structure with no change in the function. Subsequently, second mutation further shifted the position of the first mutated residue for the formation of novel ligand-specific hydrogen bond. Thus, combined effect of both mutations is essential for adoption of new function in the glucocorticoid receptor.

Both chance events and determinism play important role in the evolution of protein structure. Most proteins show parallel evolutionary trend accumulating same mutations under similar selective pressure. Thus, evolution of a protein structure appears to be predictable and deterministic. However, some mutations which do not change the protein function are not selected by the natural selection. Moreover, these mutations in future are exposed to chance events like genetic drift and ultimately shape the evolutionary trajectory. Overall, the evolution of a protein seems to be unpredictable and unrepeatable as well.

6.12 Evolution of Viruses

The recent global pandemic of coronavirus disease 2019 (COVID-19) has reinforced our interest in understanding the mechanisms involved in initiation and exponential spread of viral epidemics in human population. The evolutionary properties of viruses involved in invading a new host species and evading new vaccines are largely unknown. The emergence of novel viruses is a microevolutionary process often involving cross-species transmission. Natural selection is a powerful force in virus evolution and operates on these viruses during replication and production of progeny in the host cell. The rates of mutation and subsequent fixation of mutation in the population (i.e. substitution) of viruses are many folds higher than the cellular genes of the host species. The sexual reproduction in viruses occurs through two distinct mechanisms: recombination and reassortment. Recombination process in viruses involves coinfection of a single host cell by two viruses and generation of a hybrid molecule during replication. On the other hand, two or more segmented viruses sometimes infect a single cell and the resulting progeny virus has segments from multiple ancestries. Although recombination in a virus often leads to a new beneficial genetic combination, mutation plays a crucial role in adaptive evolution in viruses. For example, the resistance to antiviral drugs in human immunodeficiency virus (HIV) can be attributed to mutation alone rather than recombination. The intra-host population size of viruses is very large considering both the huge number of viruses per host cell and the large number of infected host cells. For instance, HIV-1 can infect 10^7 to 10^8 cells in a patient producing 10^{10} virions in a single day. Viruses lack functionless regions such as introns and pseudogenes in their noncoding genome suggesting no significant role of genetic drift in viral evolution.

RNA viruses are fast evolving pathogens and have ability to accumulate rich genetic diversity in a short period of time. The underlying evolutionary pattern can be understood using phylogenetic analysis of homologous sequences from diverse types of extant viral species or strains. Since recombination is a common phenomenon in viruses, multiple phylogenetic trees describe the complete model of a viral evolution, one tree representing each non-recombinant fragment of homologous sequences. The failure to account for recombination may not only lead to a distorted phylogenetic analysis but produce an incorrect estimation of natural selection. The recombination breakpoints can be detected using GARD program in a sequence alignment based on the phylogenetic incongruence among segments. Alternatively, RDP is a popular program to identify breakpoints in a sequence alignment using different methodologies. Natural selection acts on the RNA viruses with great efficiency due to their extremely large and well-mixed populations. Thus, positive selection is a very common mode of adaptive evolution in RNA viruses and the host genes responding to viral infections as well. The current methods of the dN/dS analysis need repeated fixation of nonsynonymous changes at a particular site in order to infer a positive selection. This type of positive selection is known as diversifying selection. In contrast, a single amino acid change in a specific lineage is known as directional selection which is hard to detect using dN/dS method. Directional selection is a rampant adaptive process in the emergence of new viruses. Most mutations especially at the nonsynonymous sites in RNA viruses like HIV are deleterious and consequently removed by the purifying selection. Although synonymous sites are generally treated as selectively neutral, synonymous sites in RNA viruses may not be selectively neutral. For instance, polymerase basic 2 (PB2) protein encoding gene is highly conserved in influenza A due to its crucial role in viral packaging. A major change in the nucleotide composition of a virus is expected when a virus jumps from one species to another species. For example, influenza A virus underwent a major change in the nucleotide composition during transmission from bird to human. Nucleotide composition of viruses can vary widely in the same families due to different hosts and vectors utilized during the transmission. There is a prevalence of epistatic interactions although mostly antagonistic in viral evolution may be due to very common use of overlapping reading frames.

Coronaviruses are zoonotic pathogens with a positive-sense and single-stranded RNA genome having origin in wild animals. The genome size of coronaviruses is extraordinarily large with size up to 32 kb. The genomic organization of all coronaviruses are similar with two large overlapping open reading frames, ORF1a and ORF1b coding for two polyproteins pp1a and pp1b, respectively. Further processing of these polyproteins produces 16 non-structural proteins (nsp1-nsp16). In addition, four types of structural proteins, namely spike, envelope, membrane and nucleoproteins, are coded by some ORFs in the remaining part of the genome. Spike proteins are key surface glycoproteins having very crucial role in interacting with distinct host cell receptors. A wide number of accessory proteins are also encoded by the genome of different viruses. Coronaviruses have lower mutation rates than other RNA viruses due to some proofreading activity of 3′-to-5′ exoribonuclease. However, this lower mutation rate in this virus is well compensated by a high rate of virus replication in hosts. In addition, coronaviruses expand their genome through frequent gene gains and losses and the genome expansion in the coronaviruses is largely contributed by their exceptionally high replication fidelity. The genome expansion has facilitated acquisition and maintenance of novel genes encoding a large number of accessory proteins performing crucial functions such as enhanced virulence, suppression of immune responses and adaptation in a specific host. Thus, the gene losses and gains in coronaviruses are key evolutionary phenomena in the emergence of novel viral phenotypes. Coronavirus genome codes for an enzyme phosphodiesterase (PDE) which is responsible for blocking the interferon-induced antiviral immune responses in the host and consequently increases the overall fitness of viruses. Interestingly, coronaviruses frequently steal some additional genes from their hosts like other viruses. For example, a viral enzyme, hemagglutinin esterase and the N-terminal domain of the spike protein are derived from cellular lectins of the host. The structural components of the coronavirus genome have received much evolutionary attention because these proteins are exposed on the viral surface to induce the immune responses in the human host. The spike proteins of coronaviruses can adapt to exploit diverse types of cellular receptors in hosts. This key adaptive feature in coronaviruses might play an important role in frequent host jump. It is still a mystery how these binding affinities of the spike proteins to cellular receptors have evolved in coronaviruses. The ongoing pandemic of novel pneumonia widely known as coronavirus disease 2019 (COVID-19) is caused by a new human betacoronavirus, SARS-CoV-2, a close relative of SARS (severe acute respiratory syndrome)-like viruses. There is about 79% similarity between SARS-CoV-2 and SARS-CoV at the nucleotide level. However, this sequence similarity is little less of about 72% in the spike protein. This virus has a zoonotic origin and appears to be more infectious than any other coronavirus known till date. Most of the closely related viruses to SARS-CoV-2 are found in bats as bats are known to be a reservoir for a wide variety of coronaviruses. The genome of this novel virus has approximately 96% similarity with the genome of a related virus (RaTG13) from horseshoe bat. In spite of very high similarity at the nucleotide level, a number of key genomic features differ between two viruses. For example, SARS-CoV-2 has a polybasic S1/S2 cleavage site insertion which might account for its high infectivity and pathogenicity in human. Interestingly, the receptor binding domain (RBD) of these two viruses has approximately 85% similarity and only one amino acid is common out of six critical amino acid residues at this domain. It appears that the receptor binding domain of SARS-CoV-2 is well optimized to bind the angiotensin converting enzyme 2 (ACE2) receptor in human like other SARS-CoV. It is possible that SARS-CoV-2 might have acquired the key mutations needed for efficient human transmission in an intermediate host like pangolins. For example, the pangolin viruses have not only amino acid sequence similarity of 97% with SARS-CoV-2 but contain all human-specific six mutations at the receptor binding domain optimized for binding ACE2 receptor. However, another possibility of acquiring some of its key mutations during a period of cryptic spread in human before its first detection cannot be excluded. There are different variants of this virus across the world (Fig. 6.7) and some amino acid sites in the spike glycoprotein accumulated not only synonymous and missense mutations but also gained few stop mutations (Fig. 6.8). We can study the genetic diversity in viruses using distinct phylogenetic clusters of genomic sequences. However, it is difficult to determine the phenotypically important mutations using genomic sequences only; it can only be validated using clinical samples in laboratories.

6.13 Exercises

1. GnRH is a neuropeptide present in the hypothalamus regulating reproductive functions. Retrieve nine orthologous coding sequences of *GnRH1* from eight primates and tree shrew using Ensembl database. Perform the multiple sequence

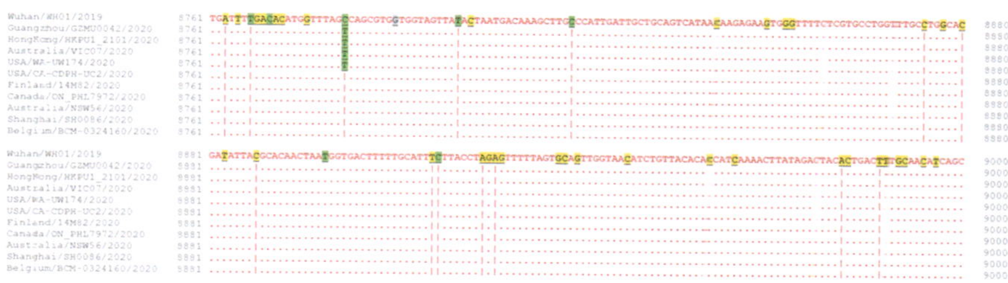

Fig. 6.7 A view of sequence alignment of 11 strains of SARS-CoV-2 on the Ensembl COVID-19 database showing variability at some nucleotide sites. The nucleotides highlighted in green and yellow colour are synonymous and missense variants

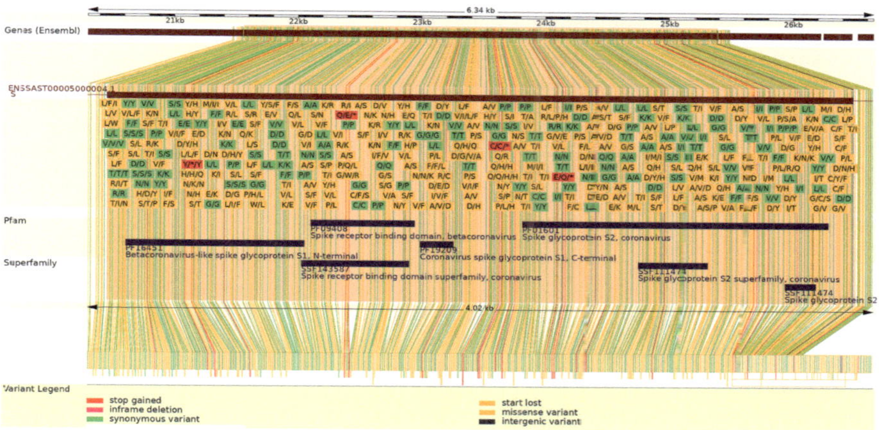

Fig. 6.8 A view of various variants of SARS-CoV-2 spike glycoprotein on the Ensembl COVID-19 database

alignment of these sequences, compute the pairwise distances between nucleotide sequences, reconstruct an unrooted neighbour joining (NJ) tree and write the tree in Newick format in the R environment. Using tree shrew an outgroup, convert the unrooted tree into rooted tree. Plot both unrooted and rooted tree in using an R package.

Solution (Figs. 6.9, 6.10, 6.11, 6.12, and 6.13)

We will use three R packages, namely "Biostrings", "msa" and "ape", for this exercise. The R commands with their output are as follows:

```
width seq                                                                 names
[1]  279 ATGAAGCCAATTCAAAAACTCCTAGCTGGCC...TGAAGAGGAAACTGGGCAGAAGAAGATTTAA Human
[2]  291 ATGTGCCTTAGAATGAAGCCAATTCAAAAAC...TGAAGAGGAAACTGGGCAGAAGAAGATTTAA Chimpanzee
[3]  279 ATGGAGCCAGTTCCAAAACTCCTAGCTGGCC...TGAAGAGGAAACCGGGCAGAGGAAGATTTAA Gibbon
[4]  291 ATGTGCCTTAGAATGAAGCCAATTCAAAAAC...TGAAGAGGAAACTGGGCAGAAGAAGATGTAA Gorilla
[5]  291 ATGTGCCTTAGAATGGAGCCAATTCCAAAAC...TGAAGAGGAAACTGGGCAGAAGAAGATTTAA Monkey
[6]  279 ATGGAGCCAATTCCAAAACTCCTAGCTGGCC...TGAAGAGGAAACTGGGCAGAAGAAGATTTAA Orangutan
[7]  279 ATGGAGCCAATTCCAAAACTTGTAGTTGGAC...TGAAGAGGAAAGTGGGCAGAAGAAGAGTTAA Lemur
[8]  279 ATGGAGCCAATTCCAAAACTCCTAGCTGGCC...TGAAGAGGAAACTGGGCAGAAGAAGATTTAA Baboon
[9]  276 ATGGAGCTGGTTCCAAAATTTCTAGCTGGAC...TGAGGAGGAAACTGGGCAGAAGAAGATTTAA Tree_shrew
```

Fig. 6.9 An overview of GnRH coding sequences under study

```
aln                                                               names
[1] ATGTGCCTTAGAATGAAGCCAATTCAAAAACTCC...ATTGAAGAGGAAACTGGGCAGAAGAAGATTTAA Chimpanzee
[2] ATGTGCCTTAGAATGAAGCCAATTCAAAAACTCC...ATTGAAGAGGAAACTGGGCAGAAGAAGATGTAA Gorilla
[3] --------------------ATGAAGCCAATTCAAAAACTCC...ATTGAAGAGGAAACTGGGCAGAAGAAGATTTAA Human
[4] --------------------ATGGAGCCAATTCCAAAACTCC...ATTGAAGAGGAAACTGGGCAGAAGAAGATTTAA Orangutan
[5] ATGTGCCTTAGAATGGAGCCAATTCCAAAACTCC...ATTGAAGAGGAAACTGGGCAGAAGAAGATTTAA Monkey
[6] --------------------ATGGAGCCAATTCCAAAACTCC...ATTGAAGAGGAAACTGGGCAGAAGAAGATTTAA Baboon
[7] --------------------ATGGAGCCAGTTCCAAAACTCC...ATTGAAGAGGAAACCGGGCAGAGGAAGATTTAA Gibbon
[8] --------------------ATGGAGCCAATTCCAAAACTTG...ATTGAAGAGGAAAGTGGGCAGAAGAAGAGTTAA Lemur
[9] --------------------ATGGAGCTGGTTCCAAAATTTC...ATTGAGGAGGAAACTGGGCAGAAGAAGATTTAA Tree_shrew
Con --------------------ATGGAGCCAATTCCAAAACTCC...ATTGAAGAGGAAACTGGGCAGAAGAAGATTTAA Consensus
```

Fig. 6.10 Multiple sequence alignment of GnRH coding sequences

	Chimpanzee	Gorilla	Human	Orangutan	Monkey	Baboon	Gibbon	Lemur
Gorilla	0.007282725							
Human	0.0072994	0.007282725						
Orangutan	0.025875581	0.025830247	0.018356642					
Monkey	0.045063387	0.044965182	0.045063387	0.033604592				
Baboon	0.049000415	0.048886771	0.049000415	0.03748058	0.00363638			
Gibbon	0.049000415	0.048886771	0.049000415	0.03748058	0.049135589	0.053119674		
Lemur	0.171899092	0.171587153	0.171899092	0.158321682	0.167454828	0.162715456	0.167454828	
Tree shrew	0.174364093	0.183784694	0.18431467	0.170457764	0.165068204	0.160237248	0.160237248	0.205209183

Fig. 6.11 Pairwise distance between sequences

```
> library(Biostrings)
> GnRH<-readDNAStringSet("GnRH.fasta")
DNAStringSet object of length 9:
```

#Fig. 6.9

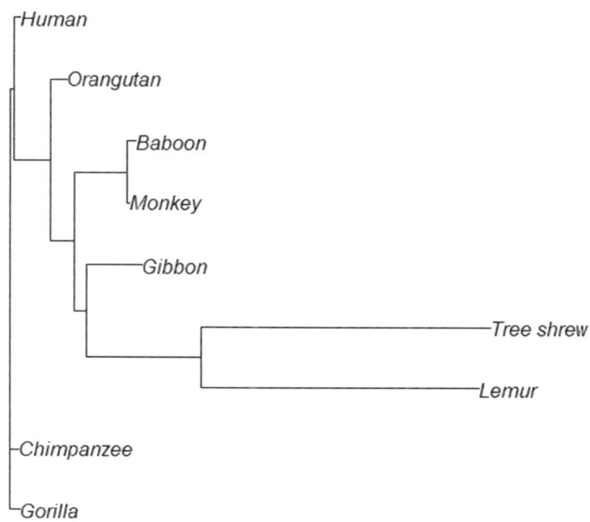

Fig. 6.12 Unrooted NJ tree

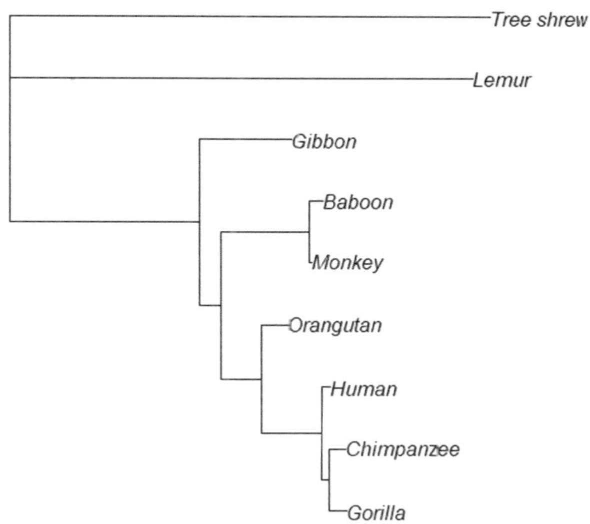

Fig. 6.13 Rooted NJ tree

```
> library(msa)
> GnRH.alignment<-msa(GnRH)
> GnRH.alignment
CLUSTAL 2.1
Call:
   msa(GnRH)
MsaDNAMultipleAlignment with 9 rows and 291 columns
```

#Fig. 6.10

	Model	df	logLik	AIC	AICw	AICc	AICcw	BIC
26	TPM1+I	18	-812.2572	1660.514	1.00E-01	1663.029	1.18E-01	1726.634
30	K81+I	18	-812.2572	1660.514	1.00E-01	1663.029	1.18E-01	1726.634
27	TPM1+G	18	-812.5733	1661.147	7.31E-02	1663.661	8.63E-02	1727.266
31	K81+G	18	-812.5733	1661.147	7.31E-02	1663.661	8.63E-02	1727.266
78	TVMe+I	20	-810.6851	1661.37	6.54E-02	1664.481	5.73E-02	1734.837
54	TIM1e+I	19	-811.8908	1661.782	5.32E-02	1664.586	5.43E-02	1731.575
79	TVMe+G	20	-810.9439	1661.888	5.05E-02	1664.999	4.42E-02	1735.354
55	TIM1e+G	19	-812.1507	1662.301	4.10E-02	1665.106	4.19E-02	1732.095
25	TPM1	17	-814.4893	1662.979	2.92E-02	1665.22	3.96E-02	1725.425
29	K81	17	-814.4893	1662.979	2.92E-02	1665.22	3.96E-02	1725.425

Fig. 6.14 Top ten best models for the GnRH coding sequences. The model (TPM1 + I) showing lowest AICc value is the best model

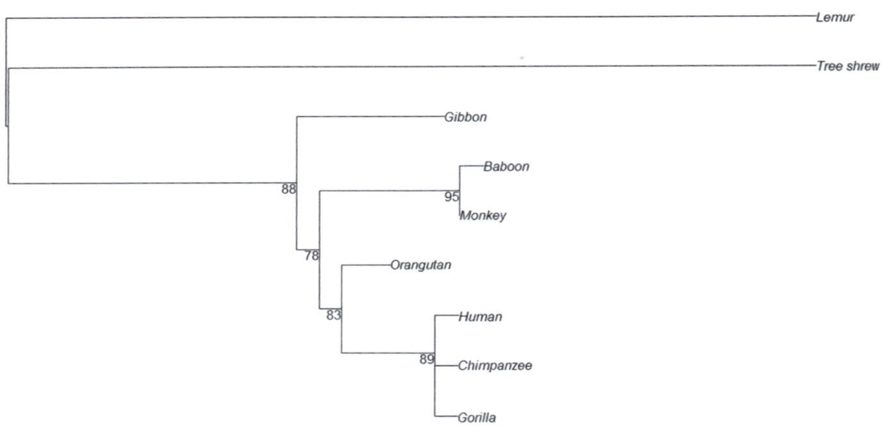

Fig. 6.15 A maximum likelihood phylogenetic tree showing bootstrap values more than 50 at respective nodes

```
> library(ape)
> GnRH<-as.DNAbin(GnRH.alignment)
> distance<-dist.dna(GnRH)
> distance
```

#Fig. 6.11

```
> tree<-nj(distance)
> write.tree(tree)
```

```
[1] "(Gorilla:0.003763678232, Chimpanzee:0.003519046866,
(((((Lemur:0.1006666062, Tree_shrew:0.1045425764):0.04148715176,
Gibbon:0.01975429486):0.004614720588, (Monkey:0.000578976943,
Baboon:0.003057402722):0.0191423532):0.008601196208, Orangutan:
0.005888531322):0.01325500637,Human:0.001950267732):0.001699432124);"
>plot(tree)
```

#Fig. 6.12

```
> rooted<-root (tree, outgroup=9)
> plot(rooted)
```

#Fig. 6.13

2. Using same set of coding sequences as in Example 1, reconstruct a maximum likelihood phylogenetic tree in the R environment in following steps:
   (a) Compute a neighbour joining initial tree.
   (b) Compute the best ten models fitting to the dataset based on AICc criteria using model test.
   (c) Optimize the initial tree using maximum likelihood method using bootstrapping.
   (d) Plot the maximum likelihood tree with bootstrap values more than 50.

Solution (Figs. 6.14 and 6.15)
We will use four R packages, namely Biostrings, msa, ape and phangorn, for this exercise.

```
> library(phangorn)
Loading required package: ape
> library(msa)
Loading required package: Biostrings
Loading required package: BiocGenerics
Loading required package: parallel

> GnRH<-readDNAStringSet("GnRH.fasta")
> GnRH.alignment<-msa(GnRH)
use default substitution matrix
> GnRH<-msaConvert(GnRH.alignment,type="phangorn::phyDat")
> distance.matrix <-dist.ml(GnRH, "F81")
> tree <- NJ (distance.matrix)
>GnRH.model <- modelTest(GnRH, tree=tree, multicore=F,model="all")
>GnRH.model[order(GnRH.model$AICc),]
```

#Fig. 6.14

```
> env <- attr(GnRH.model, "env")
> fitStart <- eval(get("TPM1+I", env), env)
> fit <- optim.pml(fitStart, rearrangement = "stochastic",
```

```
+ optGamma=FALSE, optInv=TRUE, model="TPM1")
optimize edge weights:  -812.2572 --> -812.2572
optimize rate matrix:  -812.2572 --> -812.2572
optimize invariant sites:  -812.2572 --> -812.2572
optimize edge weights:  -812.2572 --> -812.2572
optimize topology:  -812.2572 --> -812.2572
> bs =<-bootstrap.pml(fit, bs=100, optNni=TRUE, multicore=F)
> plotBS(midpoint(fit$tree), bs, p = 50, type="p")
```

#Fig. 6.15

6.14 Multiple Choice Questions

1. Which of the following statement(s) is/are true for genetic drift?
   (a) It is a stochastic process
   (b) It operates in a small population
   (c) It is a dispersive process
   (d) All the above
2. The slowest evolving protein molecule is:
   (a) Haemoglobin
   (b) Cytochrome C
   (c) Histone IV
   (d) Fibrinopeptide
3. Neutral theory of molecular evolution advocated that:
   (a) Neutral mutations are fixed by genetic drift
   (b) Nearly neutral mutations are fixed by genetic drift
   (c) Neutral mutations are fixed by natural selection
   (d) Nearly neutral mutations are fixed by natural selection
4. The genes subjected to positive selection are usually involved in:
   (a) Male reproduction
   (b) Immune response of host against pathogens
   (c) Neural development in primates
   (d) All the above
5. Most simple nucleotide substitution model is:
   (a) Jukes–Cantor model
   (b) GTR model
   (c) HKY model
   (d) Kimura two-parameter model.
6. The statistical test(s) to find best-fit evolutionary model is/are:
   (a) Likelihood ratio test
   (b) Akaike Information Criterion
   (c) Bayesian Information Criterion
   (d) All the above
7. The minimum bootstrap value indicating good confidence of a clade is:
   (a) 50%
   (b) 60%

(c) 70%
(d) 80%

8. Which of the following is not true about maximum parsimony method?
   (a) An assumption of evolutionary model is necessary
   (b) It considers only informative sites
   (c) It is prone to long branch attraction
   (d) It is prone to short branch attraction
9. The confidence of a clade in a Bayesian inference tree is indicated by:
   (a) Bootstrap
   (b) Posterior probability
   (c) Prior Probability
   (d) None of the above
10. Discarding the early phase of MCMC simulation run in Bayesian inference is known as:
   (a) burn in
   (b) burn out
   (c) burning
   (d) sampling Answer: 1. d 2. c 3. a 4. d 5. a 6. d 7. c 8. a 9. b 10. a

Summary

- The substitution rate is computed in terms of accumulation of genetic differences between individuals in an evolutionary timescale.
- Nonsynonymous mutations are subjected to selective pressure and cause phenotypic changes.
- Synonymous mutations are neutral and are fixed in a population under the impact of genetic drift.
- Neutral theory suggests that evolution is a stochastic process operating on neutral mutations where substitutions are fixed by the random genetic drift.
- Nearly neutral theory emphasized on those borderline mutations which are neither strictly neutral nor are strongly selected and called them as nearly neutral mutations.
- The higher similarity between two genes is an indicator of their being homologous (having a common ancestry).
- The best-fit evolutionary model of a particular dataset is selected using different statistical approaches such as hierarchical likelihood ratio test, Akaike Information Criterion (AIC) and Bayesian Information Criterion (BIC).
- A phylogenetic tree consists of two characteristic features, namely topology of a tree and its branch lengths.
- The reliability of a specific clade of a phylogenetic tree is evaluated using bootstrap analysis.
- In Bayesian inference of phylogeny, posterior probability distribution is usually estimated using the Markov Chain Monte Carlo (MCMC) sampling.
- The molecular clock ticks at irregular intervals and is probabilistic in nature.

- The duplication of genes and whole genomes have played an important role in the evolution of organismal complexity.
- The omega (dN/dS) is measured as the observed number of nonsynonymous substitutions per nonsynonymous site (dN) over the synonymous substitutions per synonymous site (dS).
- The omega is expected to be equal to one under the assumption of neutral evolution.
- An omega value significantly less than one indicates negative or purifying selection.
- An omega value significantly more than one indicates operation of positive or directional selection.
- The marginal stability of a protein structure is a product of neutral evolution through balance between mutation and selection.
- The spike proteins of coronaviruses can adapt to exploit diverse types of cellular receptors in hosts.

Suggested Reading

Li W-H (1997) Molecular Evolution. Sinauer Associates Inc, Sunderland

Graur D, Li W-H (2000) Fundamentals of molecular evolution. Sinauer Associates Inc, Sunderland

Nei M, Kumar S (2000) Molecular evolution and Phylogenetics. Oxford University Press, Oxford

Bromham L (2016) An introduction to molecular evolution and Phylogenetics. Oxford University Press, Oxford

Next-Generation Sequencing 7

Learning Objectives
You will be able to understand the following after reading this chapter:

- The principles of second-generation, third-generation and fourth-generation sequencing platforms.
- Next-generation sequencing data formats.
- Visualization of next-generation sequencing data.
- Methods involved in reference-based assembly and de novo assembly of short reads.
- Application of next-generation sequencing technology.

7.1 Introduction

Although Sanger sequencing technique was instrumental in completing the first human genome sequence in 2004, a worldwide effort to develop a high-throughput, cheaper and faster next-generation sequencing (NGS) technology started in the same year. This initiative has led to the development of novel next-generation technologies generating enormous number of short reads at an unprecedented speed. The first NGS technology developed in 2005 was based on the pyrosequencing method and is popularly known as the 454-genome sequencer. This technology was soon followed by two new variants of the NGS: the Solexa/Illumina sequencer and the SOLiD sequencer in 2007. Both of these new technologies could generate large number of reads in comparison to 454 but initial size of the reads generated were very short (35 bp). Next addition to the series of NGS technology was the Ion Torrent technology in 2010 based on semiconductor technology with a small instrument size at a lower cost. All these next-generation

B. K. Tiwary, *Bioinformatics and Computational Biology*,
https://doi.org/10.1007/978-981-16-4241-8_7

technologies require prior amplification of the template DNA. The necessity of prior DNA amplification was circumvented in the third-generation sequencing technology known as Pacific Biosciences (PacBio) platform. Moreover, it allows the detection of single molecules in real time and generates thousands of several kilo-base long reads suitable for de novo assembly. The latest addition to sequencing technology, Oxford nanopore is treated as the fourth-generation sequencing technology and is capable of producing ultra-long reads at a cheaper cost. Overall, the second-generation technologies have high throughput with a lower error rate and cost per base. On the other hand, the third- and fourth-generation technologies offer long read length with short running time. Thus, advantages of different generation of technologies are complementary and a hybrid-sequencing approach having a mixture of different generations of sequencing offers a better solution to whole-genome sequencing. The rapid development of next-generation technology was complemented by ever increasing computing power and development of efficient algorithms for assembly of short reads.

7.2 Sequencing Platforms

Most popular second-generation sequencing platforms are 454, Illumina, SOLiD and Ion Torrent. The PacBio and the Oxford nanopore technologies are the third-generation and the fourth-generation technologies, respectively. A comparison of different generations of sequencing platforms is briefly described in Table 7.1. The principles of each sequencing method with its advantages and limitations are described as follows:

7.2.1 454 (Pyrosequencing)

The preparation of sample starts with capturing each DNA single-stranded fragment of the DNA libraries on an agarose bead using adaptor sequences. The fragment-bead complex is secluded in water-in-oil micelles containing PCR reactants. These micelles act as a microreactor for thermal cycling (emulsion PCR) of each DNA

Table 7.1 A comparative account of different generations of sequencing platforms

Generation	Read length (bp)	No. of reads per run	Single pass error rate (%)	Time per run
First generation (Sanger ABI 3730 X 1)	600–1000	96	0.001	0.5–3 hours
Second generation (Illumina HiSeq 2500)	2 X 250	8 X 10^9	0.1	7–60 hours
Third generation (PacBio RS II)	10,000-20,000	3.5–7.5 X 10^4	13	0.5–4 hours
Fourth generation (Oxford Nanopore MinION)	10,000-20,000	1.1–4.7 X 10^4	10	48 hours

fragment into about one million copies. Finally, microreactor is broken and each 28 μm DNA-coated bead containing one DNA fragment is fitted into a well of 44 μm average diameter on a fibre-optic slide. The sequencing enzyme and primer are loaded into the well. The synthesis of complementary DNA strand is performed by passing one unlabelled nucleotide through the well at a time. The incorporation of a nucleotide leads to release of pyrophosphate with a light emission by firefly enzyme luciferase in the well. The amount of light produced by the luciferase is proportional to the number of nucleotides incorporated by the DNA polymerase. The signal for nucleotide incorporation is detected by an underlying charged couple device (CCD) sensor in a real-time basis. Finally, 400,000 short reads (400–500 bp length) are produced in parallel sequencing reactions using millions of wells. The raw reads are processed by the 454-analysis software and produce about 100 Mb data consisting of quality reads.

7.2.2 Illumina (Sequencing-by-Synthesis)

Denatured adapter-linked DNA libraries are attached at one end to the complementary oligonucleotides attached onto a flow cell. The flow cell is a closed glass microfabricated device having eight channels. The open end of the DNA fragment bends down and further hybridizes to a complementary adaptor attached on the solid surface. The DNA strands bound to the flow cell undergo solid phase PCR to produce clusters of clonal populations. This process is known as bridge amplification. The solid phase amplification process undergoes several cycles and the resulted amplified DNA is denatured to produce a cluster of around 1000 single-stranded DNA molecules. The DNA polymerase along with primers and four reversible terminator nucleotides labelled in different way are added for synthesis. The incorporation of new nucleotides at 3′ terminator base during synthesis are detected by a CCD sensor based on its colour followed by enzymatic removal of fluorophore. This cycle is repeated several times during the synthesis process. First machine using this process was known as Genome analyser (GA) and it was capable of producing paired-end reads with length up to 35 bp. Now, two new variants of the GA, HiSeq and MiSeq are being used by various laboratories with a longer read length and depth, although MiSeq machine has a lower throughput with a cheaper cost than the HiSeq machine.

7.2.3 SOLiD (Sequencing by Ligation)

The sample preparation of the SOLiD platform is highly similar to the 454 platform. The sequencing process in the SOLiD machine is performed by ligation using a DNA ligase rather than synthesis using DNA polymerase in the Illumina platform. First, the genomic DNA is sheared into 200 bp fragments and oligonucleotide adaptors are added at the both ends of the fragment. Single-molecule templates are attached to the magnetic beads and are amplified on the beads using emulsion PCR.

The beads containing one amplified DNA fragment are spread onto a glass slide surface which are subsequently loaded into a flow cell. First, a sequencing oligonucleotide primer is hybridized to an adaptor. The 5′ end of the primer ligates with an oligonucleotide hybridizing to the adjoining sequence. The ligation to the primer is subjected to competition between a combination of octamers. The ligated octamer is cleaved after detection of colour and this ligation cycle is repeated several times.

7.2.4 Ion Torrent (Semiconductor Sequencing)

The sample preparation and sequencing methodology are similar to the 454 sequencing. The sample preparation is based on the emulsion PCR, similar to sample preparation in 454 machines. The incorporation of each nucleotide is detected by a minute semiconductor pH sensor implanted in each well of the picotitre plate. The formation of phosphodiester bond during nucleotide incorporation leads to release of a proton (H^+ ion) during the polymerization process in the Ion Torrent machine instead of pyrophosphate in 454 platform. The difference in pH caused by proton release is detected by a complementary metal-oxide-semiconductor (CMOS) sensor technology.

7.2.5 PacBio

This third-generation technology has a potential to sequence single molecules with longer read length without any need for prior DNA amplification. This single-molecule real-time (SMRT) sequencing is the most widely used third-generation sequencing technology. A target double-stranded DNA molecule is ligated with hairpin adaptors at both ends to create a single-stranded closed circular DNA known as SMRTbell. The SMRTbell samples are loaded into a SMRT cell consisting of 1,50,000-minute sequencing units known as zero-mode waveguides (ZMWs). A single DNA polymerase is imbedded in the bottom of ZMW and further binds to either adaptor of the SMRTbell. The DNA polymerization process takes place in the presence of four fluorescent labelled nucleotides in the ZMW. Single fluorophore molecules are visualized near the bottom of the ZMW. The incorporation of fluorescent molecule is monitored in real time followed by termination of signal after cleavage of dye. The replication process occurring in all ZMWs of a SMRT cell is recorded in form of light pulses that are interpreted as the continuous long read (CLR). A single CLR may contain multiple passes of both strands in case DNA polymerase activity lasts longer. This sequencing process is much faster requiring a short running time up to several hours. This machine provides additional advantage of producing extremely long reads of over 10 kb suitable for de novo assembly.

7.2.6 Oxford Nanopore

Biological nanopores are widely used for the single-molecule detection and single- and double-stranded DNA sequencing. α-hemolysin is most common biological nanopore suitable for single-stranded DNA sequencing. Nanopore sequencing provides high-throughput label-free and ultra-long reads at a cheaper cost. Trans-membrane nanometre-sized pores are usually embedded in a biological membrane or in a solid-state film (Fig. 7.1). These membranes separate two compartments containing conductive electrolytes. The electrolyte ions can pass through the nanopores when subjected to a voltage difference and consequently generate signal in form of ionic current. The ionic current signal is stopped when nanopores are blocked by a negatively charged DNA molecule. The ionic current signals generated during the sequencing are segmented into discrete events. A machine learning algorithm either recurrent neural network (RNN) or hidden Markov model (HMM) transforms these segmented events to consensus sequence of template and complement strands of the double-stranded DNA molecules. The Oxford nanopore technologies have developed two nanopore sequencing platforms, namely GridION and MinION. MinION is a small USB portable device producing cheaper and faster ultra-long read (average length 10 kb) data without DNA amplification. The current MinION flow cells consist of 2048 protein nanopores arranged in 512 channels allowing parallel processing of 512 DNA molecules. It is useful in generating bacterial genome sequences in a couple of days. Unfortunately, the sequencing error rate of nanopore technology is very high in the range of 15–40% at present.

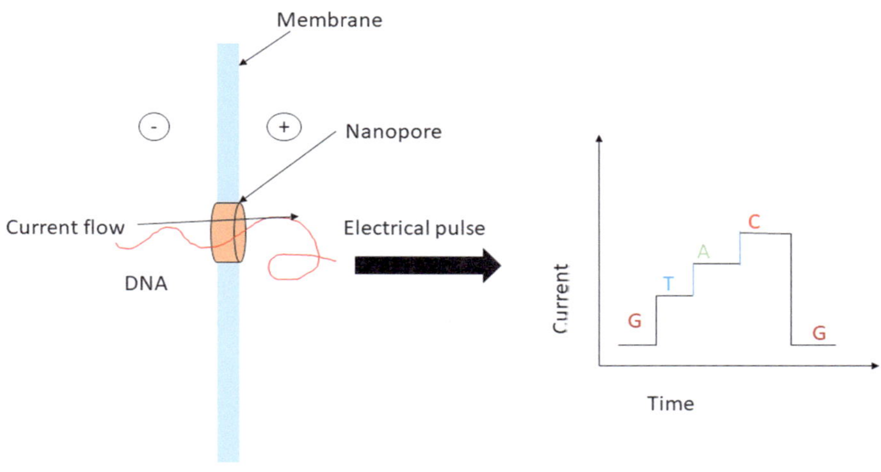

Fig. 7.1 Oxford nanopore sequencing technology is based on the changes in electrical pulses due to blocking of ionic current in a nanopore clogged by a DNA strand

7.3 Data Formats

FASTA format is a text file with a single header line starting with the character ">", which is widely used to store the nucleic acid and protein sequences. On the other hand, FASTQ format is a standard format for the NGS data with a single header line starting with the character "@". In addition, a Phred quality score for each nucleotide base is also included in the FASTQ format. The Phred quality score assigns error probability for each nucleotide base (Fig. 7.2). These error scores are the log transformed value of the error probability and computed using the formula which is as follows:

$$\text{Phred Quality score} = -10 \log_{10} \text{P, where P is the estimated error probability.}$$

Thus, a Phred score of 20 and 30 denotes an error probability of 1/100 and 1/1000, respectively. The average Phred score is used as an overall quality metric for a set of sequence. An average Phred score of 20 or more is treated as a good score for quality sequences. Moreover, high-quality sequences usually have a Phred score around 30.

SAM (Sequence Alignment/Map) format has a complex structure and is commonly used to store the reference sequence and the sequence reads aligned on the reference sequence. The SAM file has 11 mandatory fields in different columns, namely QNAME, FLAG, RNAME, POS, MAPQ, CIGAR, RNEXT, PNEXT, TLEN, SEQ and QUAL appearing in same order. All the fields are present in the SAM format even if some information is unavailable (indicated by "0" or "*"). Each field is briefly described in Table 7.2. BAM format is a binary file with compressed and indexed SAM data. The indexing is done for the BAM file in order to extract the

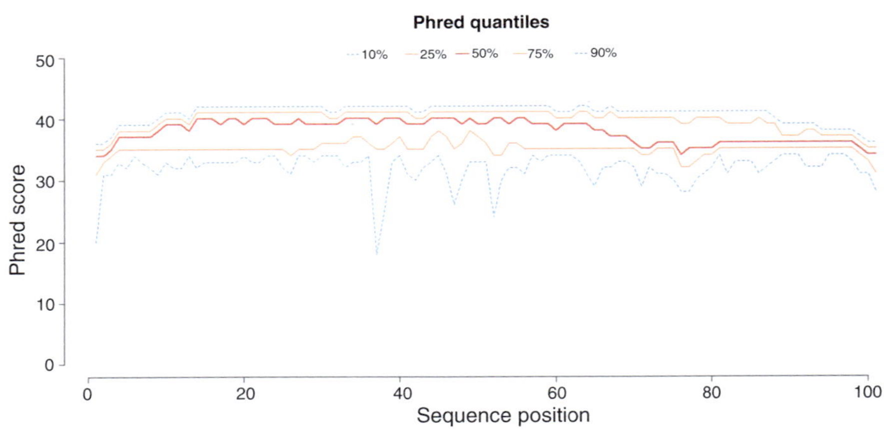

Fig. 7.2 The Phred quality score across each nucleotide base in the reads. Phred quantiles are indicated at quantiles 10%, 25%, 50%, 75% and 90% using different lines

Table 7.2 Description of various fields in the SAM format

Field	Column No.	Brief description
QNAME	1	A string for query template name. Indicated by "*" if information is not available.
FLAG	2	An integer for combination of bitwise flags
RNAME	3	Reference sequence name
POS	4	1-based leftmost position mapping of the first matching base. POS is indicated as 0 for unmapped read.
MAPQ	5	Integer indicating mapping quality computed as $-10 \log_{10} \Pr$ (mapping position is wrong)
CIGAR	6	CIGAR string
RNEXT	7	Reference name of the next/mate read
PNEXT	8	Position of the next/mate read
TLEN	9	Observed template length
SEQ	10	Segment sequence
QUAL	11	ASCII of Phred-scaled base quality+33

required information without uncompressing the whole file. SAMtools are used to convert the SAM files to the BAM files and vice versa.

7.4 Visualization of NGS Data

The UCSC Genome browser is a common web-based genome browser for visualization of NGS data. It uses two compressed binary file formats: BigBed and BigWig for transfer of voluminous NGS data. The GenomeStudio is a commercial software developed by Illumina and is capable of reading and exporting all types of NGS data formats. Mauve is a C package available for all kinds of platforms and is widely used for visualization of multiple genome sequences. The Integrative Genomics Viewer (IGV) is an open-source JAVA-based visualization software for common NGS data formats such as FASTA, FASTQ, SAM and BAM formats. Another common visualization package is the GenomeView for reading the BAM files with some advanced visualization features, for example, zooming from chromosome level to nucleotide level. The Generic Genome Browser (GBrowse) is a web-based genome visualization tool written in Perl which can be customized based on the need of the user. The preferable format for the GBrowse is GFF3.

7.5 Alignment of NGS Data

A single NGS machine is capable of producing millions of short reads. The alignment of short reads to a long reference genome sequence (reference-based assembly) or alignment of a sequence read to another sequencing read (de novo assembly) is a computationally demanding step in the overall NGS data analysis. The traditional

Table 7.3 Popular programs for next-generation sequencing data analysis

Program	Description
FastQC	A quality control JAVA-based tool for raw reads
seqTools	An R package for analysing FASTQ files
Bowtie 2	A reference-based aligner with FM-index
BWA	A reference-based aligner with FM-index
SOAP2	A reference-based aligner with FM-index
MUMmer	A reference-based aligner with suffix tree
Newbler	A de novo aligner based on OLC
Phusion	A de novo aligner based on OLC
Euler	A de novo aligner based on DBG
SOAPdenovo	A de novo aligner based on DBG
DESeq	An R package for differential gene expression analysis of RNA-seq data

methods of alignment such as Smith Waterman algorithm or BLAST have a high computational cost in terms of processing and memory usage and therefore are not suitable for alignment of short reads. There is a need for high-performance computational clusters with a high-speed storage system and large amounts of RAM for the alignment process especially in de novo assembly. The popular programs of analysis and assembly of NGS data analysis are listed in Table 7.3.

7.6 Reference-Based Assembly

The short reads are aligned to a long reference sequence using modifications of existing alignment methods such as seed-and-extend method and hash table method. In addition, new alignment method such as Burrows–Wheeler transform has been introduced for alignment of short reads. These methods are briefly described as follows:

7.6.1 Seed-and-Extend Method

The local alignment in the BLAST is based on the significant exact matches of the k-mers or seeds known as hits and comparing the matches in a hash table. The BLAST method is optimized for short read alignment by including spaced seeds in order to improve sensitivity. A spaced seed contains few internal mismatches especially at the third codon position for protein coding sequences. The alignment method consists of three steps: finding the seeds, first performing gap-free extension and rejecting the alignment with a low score and finally extension with gaps. If the seed size is very large, it is very difficult to get large exact matches and consequently the method has a low sensitivity. On the other hand, we get many random small matches with a very small size of seed increasing the running time. Moreover, this method is unsuitable for human genome with abundant repeat sequences due to lots of hits during alignment. The popular programs for alignment based on spaced seed are Short Oligonucleotide Alignment Program (SOAP), Eland, MAQ, RMAP and ZOOM.

7.6.2 Suffix Array and Suffix Tree-Based Alignment

Suffix array is a simple data structure having a list of alphabetically sorted starting positions of the suffixes in the sequence. During this alignment, a substring along a long reference sequence is searched using all suffixes starting with the substring. All these suffixes are ordered alphabetically and resulted sorted suffixes are searched for query substring using an efficient binary search. The reference string is represented by a data structure known as trie (derived from the word re<u>trie</u>val) allowing faster string matching. A trie is composed of all possible substrings representing a reference string. The terminator end of a substring is indicated by $ in suffix tries. The $ character is treated as smaller character than all other characters and is alphabetically or lexicographically first to all other characters in the substring. The lexicographic order will be $\$ < A < C < G < T$ for a DNA string. For instance, suffix arrays are created for a DNA sequence using following steps:

DNA sequence:	A	T	G	A	T	A	G	C	A	$
Position:	0	1	2	3	4	5	6	7	8	9
Suffix arrays:	9	8	5	3	0	7	2	6	4	1 (ordered alphabetically)

Suffix tree is a more complex data structure related to suffix array. A suffix tree of a string is a compressed trie for all suffixes of that string. A suffix tree is made from a string in three successive steps. First, all suffixes of a string are generated and these suffixes are treated as individual words. Finally, a compressed trie is built based on the suffixes. However, suffix array is a better space-efficient data structure than suffix tree requiring 15–20 bytes per character. On the other hand, five bytes are sufficient to represent a character in suffix array. Overall, a suffix array can store a 3.15 billion base pairs human genome in 12 GB RAM. Segemehl and Vmatch are the common alignment software based on the enhanced suffix arrays. MUMmer and OASIS are popular software using suffix tree algorithm for alignment.

7.6.3 Burrows–Wheeler Transformation (BWT)-Based Alignment

The BWT of the reference sequence is a compressed version of suffix array for compression and indexing leading to enhanced speed and reduction in the memory footprint. It is closely related to the creation of a suffix arrays. For example, T is a DNA sequence string having a length of m. We can make rotations of T by taking a character from one end of string and sticking it on the other end repetitively. As a result, a m x m matrix is created by assembling them vertically followed by sorting them alphabetically into Burrows–Wheeler matrix of T. Finally, the last column of

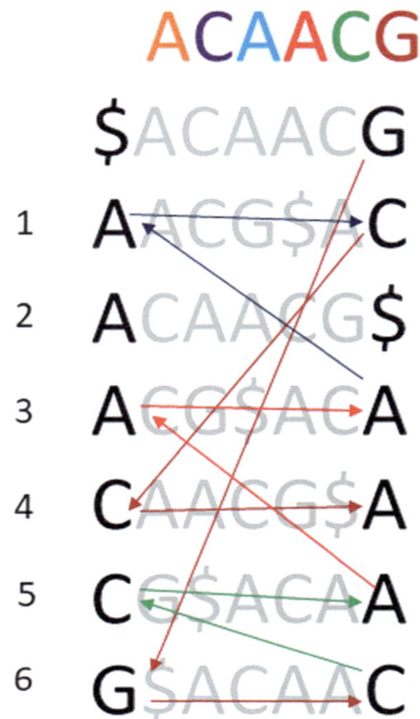

Fig. 7.3 LF mapping during Burrows–Wheeler transformation

the BWM (T) is the Burrows–Wheeler transform of T: BWT (T). For example, a DNA string ACAGCA undergoes Burrows–Wheeler transformation in following three steps.

T	**Rotations**	**BWM (T)**	**BWT (T)**
ACAGCA$ →	ACAGCA$ →	$ACAGC**A** →	AC$CGAA
	CAGCA$A	A$ACAG**C**	
	AGCA$AC	ACAGCA**$**	
	GCA$ACA	AGCA$A**C**	
	CA$ACAG	CA$ACA**G**	
	A$ACAGC	CAGCA$**A**	
	$ACAGCA	GCA$AC**A**	

The Burrows–Wheeler has an important property known as the LF (last first) mapping. The LF mapping states that the rank of a character remains same in the first column and the last column. Burrows–Wheeler transform of a string can be reversed to reconstruct T using LF mapping (Fig. 7.3). Therefore, this transformation is useful in both compression and decompression. FM-index is a data structure containing only sampled positions which are small fractions of all string positions. The sampled positions are directly retrieved using FM-index and subsequently all non-sampled positions are obtained using LF mapping. FM-index needs small memory footprints and can store a human genome with 3.15 billion base pairs in less than 3GB RAM.

The common alignment software based on the FM-index are BWA, SOAP2 and Bowtie.

7.7 De Novo Assembly

De novo assembly of short reads are performed without any prior information about the genome. This kind of assembly is necessary for novel genomes and even for a human genome with a large-scale genomic alteration induced by cancer. There are two algorithms employed for de novo assembly of short reads: overlap-layout-consensus (OLC) and de-bruijn-graph (DBG).

7.7.1 Overlap-Layout-Consensus (OLC)

The OLC algorithm is performed in three successive steps (Fig. 7.4). First, the overlapping regions between short reads are identified explicitly by aligning each read against the other read. An OLC read graph is constructed based on the layout of all the reads along with overlap information. The read graph treats each read as a node and two nodes are connected if overlap region is larger than a certain cut-off length. Thus, the total number of nodes is equal to the number of reads. The layout step involves determination of Hamiltonian path which is a NP-hard problem. Finally, the reads are aligned along the assembly path and a consensus sequence is inferred from multiple sequence alignments. This algorithm is suitable for short reads with low sequencing depth or base coverage depth (i.e. total amount of sequencing data) and is less sensitive to read errors and repeats in the genome. The sequencing depth depends upon the size of the genome and a species with large genome size needs more sequencing depth. For example, a human genome needs a minimum sequencing depth of 22. Newbler, Arachne, Phrap and Phusion are most widely used programs based on OLC algorithm. The OLC method is a better choice

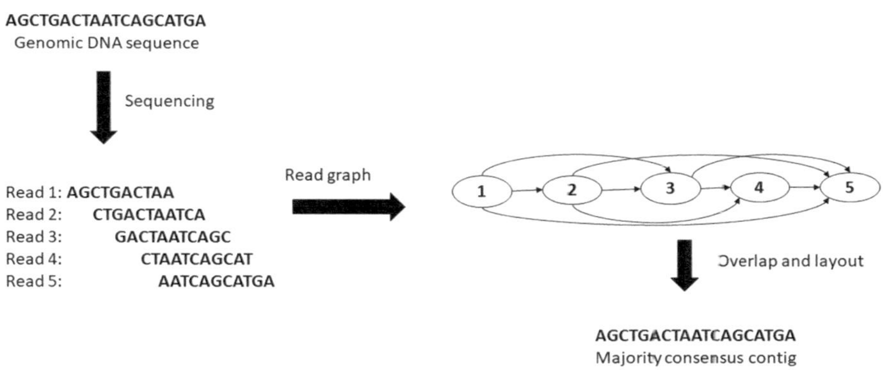

Fig. 7.4 Overlap-layout-consensus (OLC) algorithm for de novo assembly of reads

than the DBG for assembly of long and error-prone reads generated using PacBio and Oxford nanopore platforms due to its sensitive overlapping step.

7.7.2 de Bruijn-Graph (DBG)

This algorithm involves chopping of short reads into strings of specified length known as the k-mers and concomitant detection of neighbouring relationship implicitly among the k-mers (Fig. 7.5). A k-mer graph consists of individual nodes representing all unique k-mers and the edges representing exact overlap between the k-mers in the genome. Both node numbers and edge numbers are expected to be equal to the genome size and are independent of sequencing depths. However, the number of DBG nodes are likely to be much higher than the overall genome size due to inclusion of many false k-mers introduced by sequencing errors. The contig sequence is assembled from the k-mer graph through the Eulerian path using linear time algorithm which is an easier computational approach than finding Hamiltonian path. There is no need to call the consensus sequence separately in the DBG algorithm because the consensus information is already available with the k-mers. This algorithm achieves a higher memory and CPU efficiency than the OLC algorithm with a genome size with high coverage. Since the second-generation sequencing platforms usually have a high sequencing depth (>30X) in order to compensate for a short read length, the DBG is a better algorithm than the OLC for large genome assembly using short reads. However, the DBG is more sensitive to read errors and repeats unlike the OLC. Euler, Velvet and SOAPdenovo are common alignment programs developed based on the DBG algorithm.

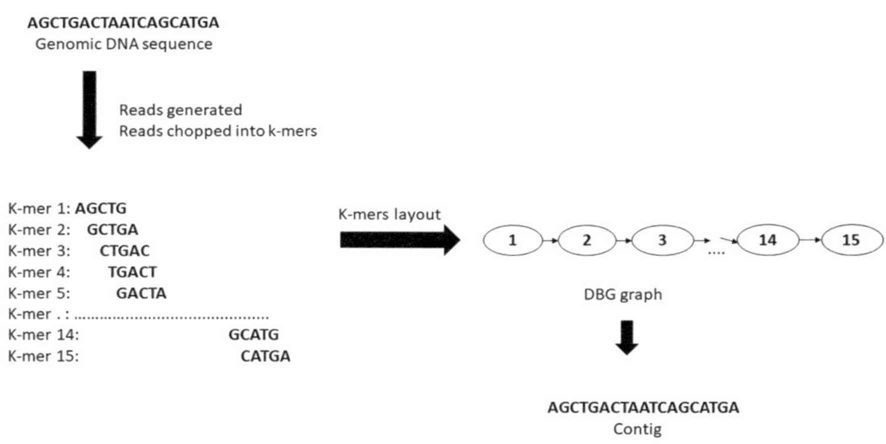

Fig. 7.5 De-Bruijn-graph (DBG) algorithm for de novo assembly of reads

7.8 Scaffolding

Short reads are assembled into large number of long and accurate contigs using both OLC and DBG algorithms. Scaffolding is process of ordering and orienting contigs similar to their position in the genome. Scaffold is a collection of contigs positioned in a linear order and correct orientation. However, there may be some gaps between contigs representing unknown sequences. This process is performed in four successive steps: contig orientation, contig ordering, contig distancing and gap closing. The scaffolding problem is approximately solved using graph theory where each contig is a node and linking read pairs are treated as edges. The read pairs are generated during paired-end sequencing and also known as *mates*. Mates may overlap with each other in the middle of the fragment based on their lengths. Some of the mates at one end of a contig are likely to pair with the mates in another neighbouring contig and are known as spanning pairs. The presence of spanning pairs indicates the proximity of two contigs. The distance between the contigs is estimated from the expected distance between the linking reads obtained from the fragment length distribution. This distance is included in the scaffolding graph as lengths to the edges. Finally, one scaffold per chromosome of the genome including gaps of correct lengths between the contigs is obtained. Some of the most popular contemporary scaffolders are Bambus, Opera, ABySS, SGA and SOAPdenovo2.

7.9 Applications of Next-Generation Sequencing

There is a variety of applications of next-generation sequencing in genomics. Some of the areas of applications are as follows:

7.9.1 Whole-Genome Sequencing

Whole-genome sequencing is useful for understanding the genome of novel organisms. The short reads from a novel genome is assembled using de novo assembly methods. The genetic variations present in the regulatory sequences and non-coding regions have been implicated in genetic susceptibility to diseases. These variations along with variations in the protein coding sequences are determined using whole-genome sequencing. However, the whole-genome sequencing incurs about ten-fold more cost in comparison to exome sequencing. The genetic variation present in a species can be unveiled and measured using existing reference genome of that species. This method helps us in understanding the single nucleotide polymorphism (SNP) and copy number variations (CNV) in relation to the biology of genetic diseases and developing potential therapeutic drugs. The causative variants of cancer, cardiovascular diseases and neurological diseases have been successfully determined using whole-genome sequencing. There are several ongoing whole-genome sequencing projects across the world such as Human Genome Project-Write, 100,000 genome project and GenomeAsia 100 K (GA 100 K).

7.9.2 Transcriptomics

The transcriptome refers to complete set of all RNA molecules encoded by the active genes present in a cell. Thus, the transcriptome of a cell is likely to change with respect to its functional and temporal state. The transcriptome of a cell can be analysed for gene expression studies and identify all upregulated and downregulated genes under a particular condition. Although microarray technology which is based on hybridization is routinely used for global gene expression studies on thousands of target genes, NGS technology is technically superior to microarray technology because it can detect novel transcripts and splice-events. Moreover, it is free from unspecific hybridization. The RNA-seq is an RNA-specific next-generation sequencing protocol to analyse the RNA sequences and their expression levels without any knowledge about the target genes. It can detect the accurate gene expression levels quantitatively and identify tissue-specific transcript isoforms as well. The exact location of exon boundaries and polymorphism in the sequences are also detected during RNA-seq experiment. Complementary DNA is synthesized from RNA extracted from the cell. RNA sequencing allows us to understand not only differential gene expression but also mutation, gene fusion and splicing of RNA. Moreover, it provides us additional information regarding non-coding RNAs such as microRNA (miRNA) and long non-coding RNA (ncRNA) linked to the pathogenesis of various diseases. The abundance of a transcript is measured in terms of read counts or RPKM (reads per thousand nucleotides in transcripts per million reads) or FPKM (fragment per thousand nucleotides per million map reads) or TPM (Transcripts per million). The RPKM and FPKM are applicable for single-end and paired-end RNA-seq, respectively. Thus, FPKM takes into account the fact that two reads can map a single fragment. TPM measures the frequency of a transcript within a sample.

7.9.3 Epigenomics

Epigenetics is concerned with DNA modification through methylation or hydroxymethylation affecting both DNA and associated histone proteins in the nucleosome. A high-resolution genome-wide epigenomic profile can be developed using next-generation sequencing platforms. The genome-wide epigenetic modifications between the healthy and the disease status are compared. The N-terminal ends of histones are usually modified under environmental or developmental signals and this post-translational genome-wide modification can be mapped using chromatin immunoprecipitation (ChIP). The histone-modification profiling includes cross-linking DNA to histones using formaldehyde, digesting the cross-linked DNA followed by immunoprecipitation using appropriate antibodies raised against a particular post-translational N-tail modification. Finally, the DNA fragments associated with the histone proteins are purified and purified DNA is sequenced. This method is known as chromatin immunoprecipitation sequencing (ChIP-seq). Alternatively, chromatin immunoprecipitation-exonuclease (ChIP-exo) method involves digestion of the precipitated DNA fragment at either end using exonuclease. The most dominant sequencing platform used for ChIP-seq is Illumina platform followed by SOLID platform. On the other hand, a wide range of methods such as methylation-dependent enzymatic restriction, methyl DNA enrichment and direct bisulphite conversion are used for profiling DNA methylation. Moreover, PacBio and Oxford nanopore platforms can directly detect DNA methylation at the level of single molecule.

7.9.4 Exome Sequencing

The whole-exome analysis is a cost-effective approach to determine the genome-wide variations. The whole-exome sequencing involves the sequencing of all protein encoding genes in the human genome in order to identify all disease-causing variants. Only targeted regions of the genome are sequenced in exome sequencing with a lesser cost than the whole-genome sequencing. The sequencing data is processed in three successive steps: read quality control, read mapping and post mapping control. The first step involves removal of all low-quality reads followed by mapping of remaining reads to the reference genome. The post-mapping processing includes measuring mapping statistics, merging BAM files, marking duplicates and recalibration of base quality scores. The exome sequencing has been used successfully to identify many novel causal genes associated with various diseases. Moreover, exome sequencing can detect some variants often missed by the low-coverage whole-genome sequencing.

7.10 Examples

1. Construct a suffix array for a DNA string “ATCGATAACGA”.

Solution
The suffixes of the DNA string are as follows:

```
[0] ATCGATAACGA
[1] TCGATAACGA
[2] CGATAACGA
[3] GATAACGA
[4] ATAACGA
[5] TAACGA
[6] AACGA
[7] ACGA
[8] CGA
[9] GA
[10] A
```

All suffixes of the DNA string are then ordered alphabetically:

```
[10] A
[6]  AACGA
[7]  ACGA
[4]  ATAACGA
[0]  ATCGATAACGA
[8]  CGA
[2]  CGATAACGA
[9]  GA
[3]  GATAACGA
[5]  TAACGA
[1]  TCGATAACGA
```

2. Construct a Burrows–Wheeler Transform of a DNA string "AACTGACAGCAT AAGCAT".

Solution

```
DNA string:
AACTGACAGCATAAGCAT$
Rotations of the string:
AACTGACAGCATAAGCAT$
ACTGACAGCATAAGCAT$A
CTGACAGCATAAGCAT$AA
TGACAGCATAAGCAT$AAC
GACAGCATAAGCAT$AACT
ACAGCATAAGCAT$AACTG
CAGCATAAGCAT$AACTGA
AGCATAAGCAT$AACTGAC
GCATAAGCAT$AACTGACA
CATAAGCAT$AACTGACAG
ATAAGCAT$AACTGACAGC
TAAGCAT$AACTGACAGCA
AAGCAT$AACTGACAGCAT
AGCAT$AACTGACAGCATA
GCAT$AACTGACAGCATAA
CAT$AACTGACAGCATAAG
AT$AACTGACAGCATAAGC
T$AACTGACAGCATAAGCA
$AACTGACAGCATAAGCAT

Sorted Rotations:
$AACTGACAGCATAAGCAT
AACTGACAGCATAAGCAT$
```

```
AAGCAT$AACTGACAGCAT
ACAGCATAAGCAT$AACTG
ACTGACAGCATAAGCAT$A
AGCAT$AACTGACAGCATA
AGCATAAGCAT$AACTGAC
AT$AACTGACAGCATAAGC
ATAAGCAT$AACTGACAGC
CAGCATAAGCAT$AACTGA
CAT$AACTGACAGCATAAG
CATAAGCAT$AACTGACAG
CTGACAGCATAAGCAT$AA
GACAGCATAAGCAT$AACT
GCAT$AACTGACAGCATAA
GCATAAGCAT$AACTGACA
T$AACTGACAGCATAAGCA
TAAGCAT$AACTGACAGCA
TGACAGCATAAGCAT$AAC

Transform of the DNA string:
T$TGAACCCAGGATAAAAC
```

7.11 Multiple Choice Questions

1. The solid phase PCR in a flow cell is known as:
   (a) Emulsion PCR
   (b) RT-PCR
   (c) Bridge amplification
   (d) Multiplex PCR
2. In the PacBio platform, DNA polymerization occurs in:
   (a) Flow cell
   (b) ZMW
   (c) Microreactor
   (d) Well
3. The average length of reads produced by the MinION is:
   (a) 1 kb
   (b) 5 kb
   (c) 10 kb
   (d) 20 kb.
4. The Phred score of high-quality reads is 30 and it indicates an error probability of:
   (a) 1/100
   (b) 1/1000
   (c) 1/10000
   (d) 1/100000.

5. A SAM file is converted to a BAM file and vice versa using:
   (a) SAMtools
   (b) BAMtools
   (c) CONVERTtools
   (d) None of the above
6. In a suffix trie, the terminator end of a DNA substring contains:
   (a) $
   (b) A
   (c) C
   (d) T
7. LF mapping is a property of:
   (a) Suffix array
   (b) Suffix tree
   (c) Burrows–Wheeler transformation
   (d) Spaced seed
8. The layout step of the OLC algorithm is based on:
   (a) Eulerian path
   (b) Hamiltonian path
   (c) Shortest path
   (d) Longest path
9. The k-mer graph in a DBG algorithm is assembled into contig using:
   (a) Eulerian path
   (b) Hamiltonian path
   (c) Shortest path
   (d) Longest path
10. The abundance of a transcript in RNA-seq data is measured using:
   (a) RPKM
   (b) FPKM
   (c) TPM
   (d) All the above

Answer: 1. c 2. b 3. c 4. b 5. a 6. a 7. c 8. b 9. a 10. d

Summary

- The PacBio technology has a potential to sequence single molecules with longer read length without any need for prior DNA amplification.
- FASTQ format is a standard format for the next-generation sequencing data including a Phred quality score for each nucleotide base.
- BAM format is a binary file with compressed and indexed SAM data.
- Suffix array is a simple data structure having a list of alphabetically sorted starting positions of the suffixes in the sequence.
- The Burrows–Wheeler transformation (BWT) of the reference sequence is a compressed version of suffix array for compression and indexing.

- FM-index is a data structure containing only sampled positions which are small fractions of all string positions.
- The OLC method is a better choice than the DBG for assembly of long and error-prone reads generated using PacBio and Oxford-nanopore platforms.
- The DBG is a better algorithm than the OLC for large genome assembly using short reads.
- Scaffold is a collection of contigs positioned in a linear order and correct orientation.
- RNA-seq detects the accurate gene expression levels quantitatively and identifies tissue-specific transcript isoforms.
- The abundance of a transcript is measured in terms of read counts or RPKM or FPKM or TPM.
- The whole exome sequencing involves the sequencing of all protein encoding genes in the human genome.

Suggested Reading

Ye SQ (2016) Big data analysis for bioinformatics and biomedical discoveries. CRC Press, Boca Raton

Brown SM (2013) Next generation DNA sequencing informatics. Cold Spring Harbor Laboratory Press, Cold Spring Harbor

Janitz M (2008) Next generation genome sequencing: towards personalized medicine. Wiley-Blackwell, Somerset

Masoudi-Nejad A, Narimani Z, Hosseinkhan N (2013) Next generation sequencing and sequence assembly: methodologies and algorithms. Springer, New York

Systems Biology

8

Learning Objectives

You will be able to understand the following after reading this chapter:

- Definition of systems biology and complex biological systems.
- Description of various types of computational models.
- Common motifs present in biological systems.
- The mechanistic basis of robustness in biological systems.

8.1 Introduction

The molecular biology has provided us the finer details of the structure and function of genes and proteins in a cell. As a result, we can explain the molecular basis of many biological processes such as reproduction, development and initiation of a disease. The whole-genome sequencing of human and other organisms has contributed significantly in understanding the molecular basis of a biological process with a better understanding of genes and proteins involved in the process. However, this molecular level understanding is not sufficient for systems-level understanding of the biological systems. Systems biology is an emerging approach to understand the holistic biology at the systems level. Here, we study the components of a system and their inter-relationship, behaviour of a system on temporal and spatial scales and the properties controlling the behaviour of a system. There are four fundamental steps in systems biology. First, we identify the components of a system and model the system. Then, we perturb the system and monitor the changes in the behaviour of the system. Finally, we refine and test the model in iterative manner under different conditions. Thus, systems biology model provides new insights into the mechanistic link between genotype and phenotype.

B. K. Tiwary, *Bioinformatics and Computational Biology*,
https://doi.org/10.1007/978-981-16-4241-8_8

8.2 Complex Biological System

Complex biological systems consist of simple components such as genes, proteins and cells but exhibit complex dynamic behaviour at different scales of biological organization. This property is known as the emergent property of a system and this phenomenon is called as emergence. The components of a biological system are highly heterogeneous. Although all organisms are made of cells, the cells are diverse in structure and function across the organisms. Even, the behaviour of each cell cannot be approximated as an average behaviour of whole system. Each component of the system shows complexity in their structure and function. For example, each protein has its own primary, secondary and tertiary structure with functional dynamics in a cell. The interaction among the components in a complex system is highly selective. This selective nature of interactions finally leads to a stable complex macromolecular structure and maintenance of functional stability. Although evolution is a stochastic process, the design principles in a complex biological system are not random. Natural selection has preferred some design principles that seem to be optimal solutions under a given environment.

8.3 Computational Model

Computational model is a simple representation of a real complex system having only essential aspects of the system. The computational models are generally defined based on the question in mind. For example, how can we engineer a metabolic pathway in a yeast cell to maximize the production of ethanol? For building a computational model, model structure of a system is first specified in the form of mathematical equations and the set of parameters under biologically feasible variations are also identified. Large complex models are generally more accurate due to a large number of parameters. However, many parameters are unknown in the model and numerical simulation of large models is very hard. Therefore, a trade-off between simplicity and accuracy needs to be maintained in computational modelling. Thus, a good model only captures the fundamental features of a system ignoring other irrelevant details. Overall, a good computational model should be simple, accurate, reusable and capable to predict the behaviour of a system.

8.4 Network Models

We need to understand first the interactions between components of a system in order to fully understand the overall behaviour of a complex system. A biological network consists of several nodes (N) represented by various genes/proteins/metabolites in a system and connected by edges (L) showing interaction between two and more nodes. Highly connected nodes in a network are known as the hubs. Transcription regulatory network, protein–protein interaction network, signalling network, neural network and metabolic network are different types of biological

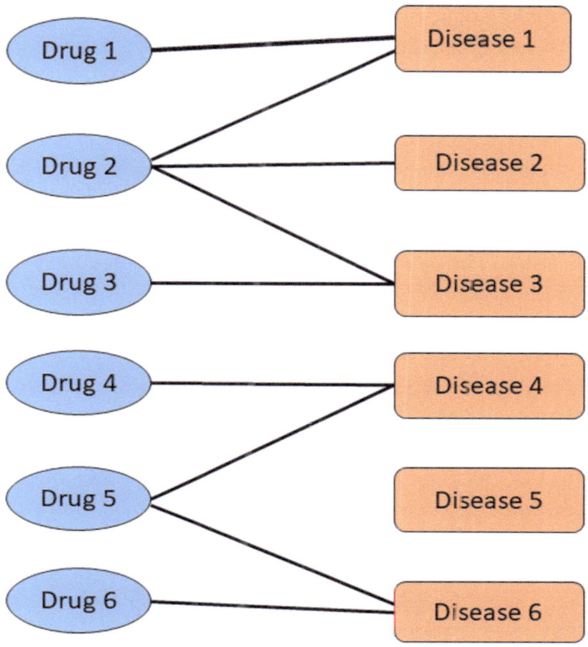

Fig. 8.1 A bipartite network showing connections between two sets of nodes: diseases and drugs

networks. The number of nodes and edges varies widely in these biological networks. For example, there are 297 nodes (neurons) and 2345 edges (synapses) in the neural network of small worm *C. elegans*. On other hand, a human brain is composed of 10^{11} neurons and each neuron is connected to 7000 synapses on average.

A biological network is mathematically represented by a graph. The interaction between two nodes in some cases represented by a directed edge when there is a particular direction in the interaction. All nodes are connected to each other in a complete graph. However, the interaction may be two sided in many cases represented by undirected edges between two nodes. The most important property of a node in a network is its degree (k) (i.e. the number of links to other nodes). The average degree ($<k>$) of a network is a global property reflecting the overall connections between nodes. The overall connection in the network is mathematically represented in the form of an adjacency matrix having two elements 0 and 1 indicating either absence or presence of an interaction. The adjacency matrix of an undirected network is symmetric and is typically sparse in case of real networks. Sometimes, a network consists of two disjoint sets U and V where node of a particular set can connect to the node of other set. This kind of network is known as bipartite network and best exemplified by a disease-drug network (Fig. 8.1). In a network, the distance between two nodes is measured by the path length where a path is the total number of links between two nodes. The geodesic path (d) is the shortest path (i.e. path with minimum number of links) between two nodes. The diameter (d_{max}) is the longest geodesic path between two connected nodes in a network. The average path length (<d>) of a network is the average of shortest paths between all pairs of nodes.

Clustering coefficient is the measurement of well-connected clusters in the network. The local clustering coefficient (C) is a measure of local density which indicates the degree to which the neighbours of a particular node are connected to each other. Moreover, the global density of the whole network is represented by the average clustering coefficient (<C>) taking the average value of all nodes in the network. The clustering coefficient of small degree nodes is significantly higher than the clustering coefficient of the hubs. Thus, the local neighbourhood of a hub node and a small degree node is sparse and dense, respectively.

The concept of centrality is based on detection of most important node in the network. The degree of node is the simplest measure of centrality and commonly known as degree centrality. The eigenvector centrality is high for nodes having connections with other important nodes. Between centrality measures the frequent position of a node between the paths of other nodes in the network. Closeness centrality indicates fast communication of a node with other nodes. The graph density is an indicator of number of connections in a subset of nodes. Most of the biological networks are not dense but found to be sparse with graph density less than 0.1. Clique is a sub-graph where all nodes are connected to all other nodes and a clique has a clustering coefficient of 1.

There are two network models proposed to represent the real properties of biological networks: random network and scale-free network. A random network G (N, p) is defined as the network with N nodes connecting with each other randomly with probability p (Fig. 8.2). When we increase the value of p from 0 to 1, the network increasingly becomes denser with linear increase in average number of edges. The random network model is also known as the Erdos and Renyi model in honour of their valuable contributions to this theory. Most nodes in this model are likely to have average degree and the degree distribution approximates Poisson distribution in a random network. The distance between two random nodes in a network is unexpectedly small. This is known as small-world phenomena. The average distance between two random nodes <d> is directly proportional to log N/ log <k>. Although real networks are not random, the random model is used as a reference guide to explain the characteristic features of real networks.

The degree distribution of real biological networks is approximated by a power law degree distribution which is as follows:

$$P_k \sim k^{-\gamma}$$

The exponent γ is known as the exponent of power law equation. The overall behaviour of a scale-free network depends on the value of this exponent. The networks with a power law degree distribution are known as scale-free networks (Fig. 8.3). Power law distribution has a higher peak and fat tail with a characteristic straight slope on a log-log scale (Fig. 8.4). The probability of finding a high degree or hub node is many folds higher in a scale-free network than a random network. The

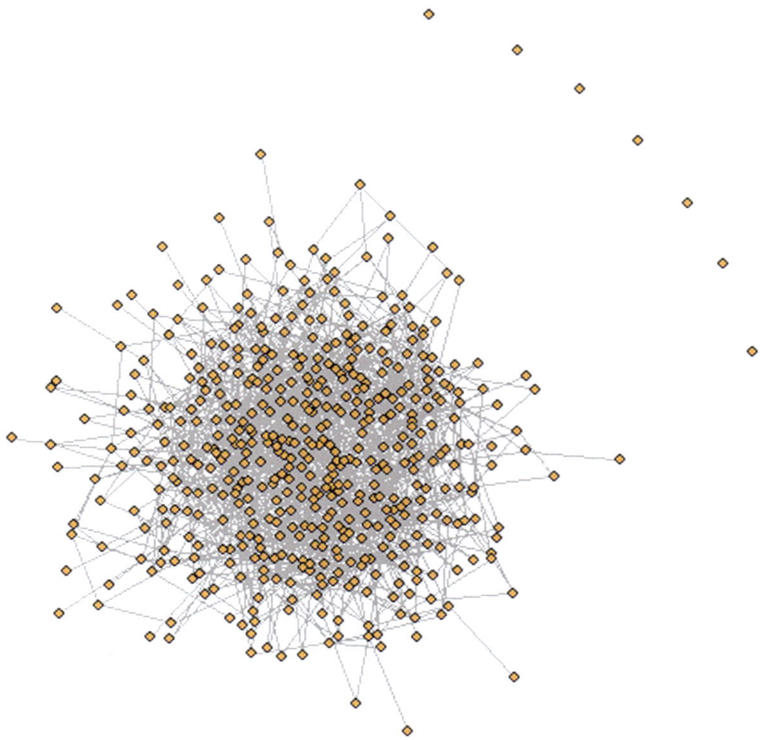

Fig. 8.2 A random network consisting of 500 nodes based on Erdos–Renyi model showing random connections between nodes

hubs have a tendency to grow polynomially and consequently occupy a large size in large networks. The value of the exponent γ varies from 2–3 in a scale-free biological network such as the protein interaction network. The scale-free networks are usually ultra-small due to the presence of hubs connecting numerous small nodes. However, the networks show the small-world property and look like a random network in case the value of exponent γ exceeds 3. On the other hand, the value of exponent less than 1 in a scale-free network is not expected unless hubs have many self-loops and there are multiple connections between same pair of nodes. Hubs in a biological network prefer to connect to nodes with less degrees rather than connecting other hubs showing disassortative nature of biological network. Biological network consists of modules and each module is characterized by a group of physically or functionally connected nodes performing a particular function.

The interaction between coexpressing genes may be inferred using quantitative measures such as correlation coefficient and mutual information approach on global gene expression data. The correlation coefficient measures linear relationship between two variables, whereas mutual information is a measure of non-linear and non-continuous dependency among two variables. Weighted correlation network

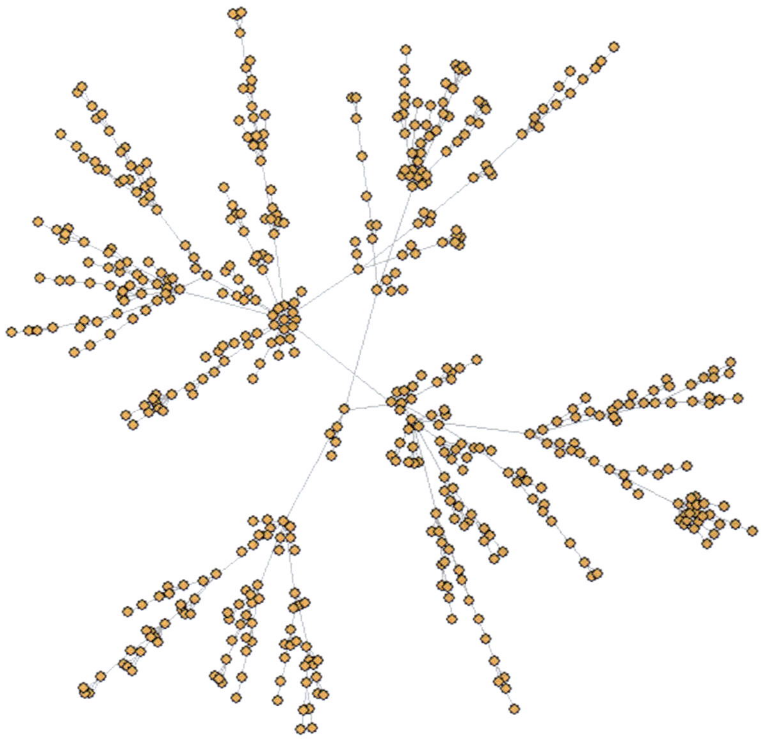

Fig. 8.3 A scale-free network consisting of 500 nodes based on Barabasi–Albert model showing hub nodes with many connections

analysis (WGCNA) is popular program for inferring correlation network among highly correlated genes from gene expression data and to identify significant modules in the network. Moreover, this method is useful in detecting correlation between image data, genetic marker data, proteomics data and other high-throughput data. Mutual information is an information theoretic approach which captures those interactions which are non-linear in nature. When the mutual information between a pair of nodes exceeds a given threshold, the nodes are connected by an edge. All pairwise connections between nodes result in a coexpression network. There are various algorithms such as CLR, ARACNE and MRNET based on mutual information approach. All these algorithms are available in a package known as *minet* in the R language and environment.

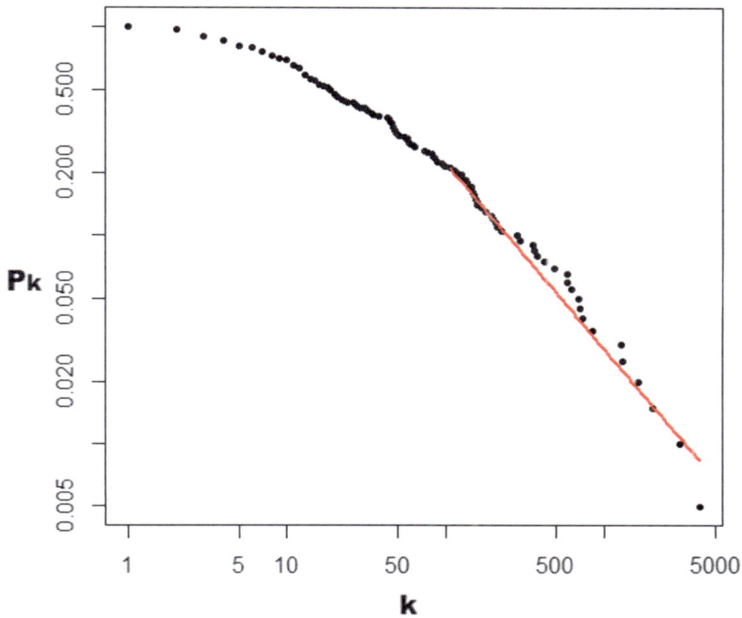

Fig. 8.4 The degree distribution of scale-free network on a log-log scale showing characteristic straight slope. k is the degree of a node in log scale and P_k represents probability of degree in a log scale. Red straight line indicates the power law fit

8.5 Genome-Scale Metabolic Model (GSMM)

The metabolic function of a cell can be represented by a mathematical relationship between metabolites, reactions, enzymes and the encoding genes. This is popularly known as gene-protein-reaction (GPR) relationship. This kind of model evaluates the complex interactions between genotype, phenotype and environment. The mathematical framework of GSMM is based on two fundamental assumptions. The first assumption is the conservation of total charge and mass in the system (i.e. total metabolites produced equal the total metabolites consumed). The second assumption is that the system maintains steady state (i.e. the concentration of internal metabolites does not change with respect to time). *Flux balance analysis* is the most common method to estimate the flow of metabolites in a metabolic network. It helps in the prediction of growth rate of an organism or rate of production of a useful metabolite. All metabolites are balanced in a cellular system by a combination of enzymes. This balanced and stable metabolite flux is known as *flux mode*. The values of metabolic fluxes are determined in a typical constraint-based model. Sometimes, thermodynamic constraints are also added to the model in addition to steady-state constraint to reduce the solution space. A set of *elementary modes (EM)* are all possible routes (pathways) in cellular metabolic network that convert some reactants to some

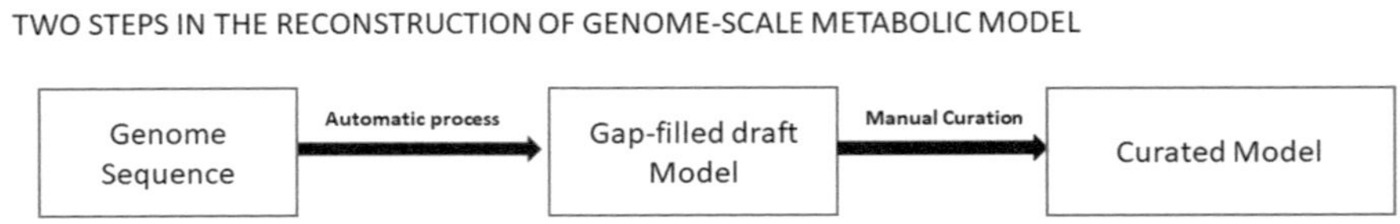

Fig. 8.5 Reconstruction of a genome-scale metabolic model

products. The total number of EMs in the cellular network indicates the overall redundancy and robustness to perform a certain function. *Extreme pathways* are the subsets of elementary modes. When both internal and external reactions are irreversible in a metabolic network, a set of elementary modes becomes identical to a set of extreme pathways. A GSMM of an organism is built in two major steps (Fig. 8.5). The first step is automatic reconstruction of a gap-filled draft model based on the genome annotation of a species. The second step is the manual refinement of the draft model using available experimental biochemical data. The building of a GSMM starts with downloading the genome sequence of a particular species. Then, the genes are annotated using bidirectional BLAST using high matching length of the query sequence (70%), moderate amino acid sequence identity (40%) and a very low e-value ($<1 \times 10^{-30}$) in the first model. In the second model, KEGG Automatic Annotation server (KAAS) is used for functional annotation of all amino acid query sequences. A draft model is developed based on both annotated models. This draft model is further refined using biochemical information from public databases such as KEGG (filling of missing genes), MetaCyc (direction of reaction), CELLO (position subsystems) and TCDB (transport reaction). Subsequently, manual corrections like gap filling, deletion of error reactions, a check for mass-charge balance and addition of species-specific information are performed in sequential phases. Further, a software environment like COBRA Toolbox or sybil package in R simulates the SBML (System Biology Markup Language) model with flux balance analysis (FBA) using a desired objective function such as growth rate or product formation (Fig. 8.6). First, all metabolic reactions are mathematically represented in the form of a stoichiometric matrix having stoichiometric coefficient of each reaction. Each reaction is defined by a minimum allowable flux (lower bound) and maximum allowable flux (upper bound). The second step is to define a phenotype in form of an objective function. For example, if the objective function is biomass production (i.e. the conversion of metabolites into biomass), an artificial biomass reaction based on experimental measures is added as an extra column in the stoichiometric matrix. As a result, we can predict the maximum growth rate of an organism by computing the conditions allowing maximum flux through the biomass reaction. Thus, objective function is a quantitative indicator of relative contribution of each reaction to the phenotype. The optimization problem is solved in flux balance analysis using linear programming where an *objective function* (Z) is either minimized or maximized. The objective function (Z) is computed by multiplying each reaction flux (v_i) with known constant (c_i) and then adding all the values. FBA finds a solution for steady state ($Sv = 0$) where each reaction is bound by an upper

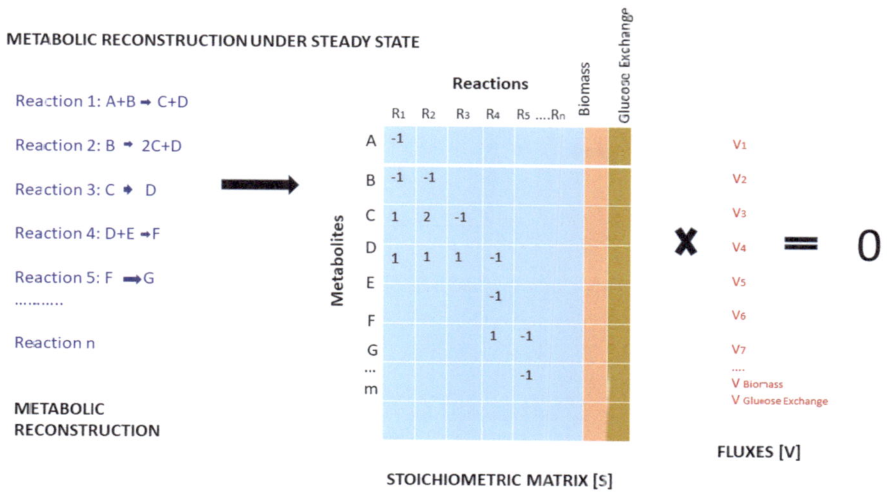

Fig. 8.6 The metabolic reconstruction has a list of stoichiometrically balanced biochemical reactions. This reconstruction is converted into a stoichiometric matrix of size m x n, where each row and column represent metabolite and reaction, respectively. Here, two additional reactions are added into the matrix to represent growth (biomass reaction) as objective function and exchange reaction of glucose inside and outside the cell. The V_{Biomass} is the objective function to be maximized during optimization. The flux through each reaction under steady state is SV = 0 which is defined by a system of linear equations

bound value and a lower bound value. A solution space is defined by the set of all possible points of an optimization problem satisfied by the mass balanced constraints having an upper and a lower bound for each reaction. The optimization for a biological objective function such as ATP utilization or biomass production finds an optimal flux within the solution space (Fig. 8.7). The flux value obtained by the FBA is not a unique optimal solution. A biological system can achieve same objective value using alternative pathway. The identification of alternate pathways using FBA is known as flux variability analysis (FVA) which measures maximum and minimum possible fluxes through a particular reaction constrained by keeping the objective function at its optimal value. Finally, the model is validated by comparing in silico result with experimentally observed phenotype. All GSSM models have some missing reactions and thus are incomplete due to our knowledge gaps.

Since all enzymes are not active in each type of a cell or under a variable culture condition, specific algorithms are necessary to simulate a particular condition in silico. Sometimes biological, physical or chemical constraints are derived from experimental data obtained under a particular condition to build a biological meaningful model known as context-specific or condition-specific GSMM. Here, the omics (i.e. transcriptomics, proteomics and metabolomics) data under a particular condition is mapped onto the template framework of a general purpose model. There are specific algorithms to reconstruct a cell or a strain specific model (Table 8.1).

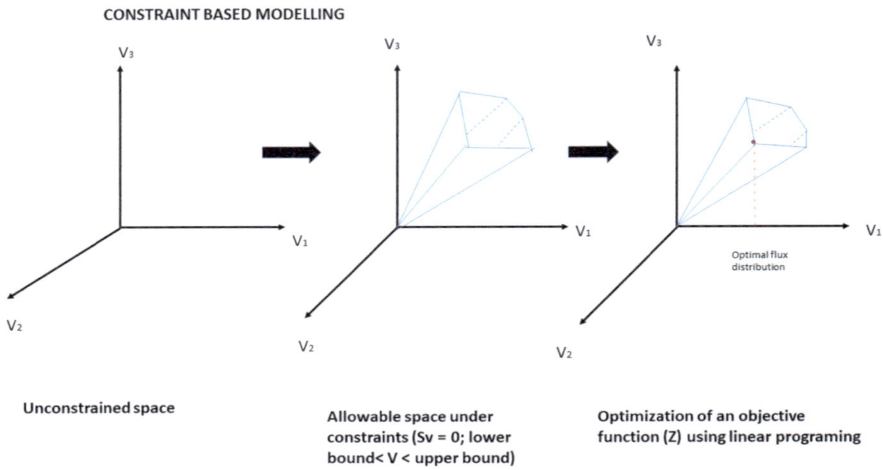

Fig. 8.7 The flux distribution of a metabolic network may lie at any point in an unconstrained space. After defining constraints in the model, any flux distribution is feasible only within the allowable space. Further, the objective function is optimized using linear programming to find a single optimal flux distribution lying on the edge of the allowable space

Table 8.1 Algorithms used for context-specific genome-scale models

Method	Description of the algorithm
MBA	Any omics data can be used to identify high and medium confidence reactions. Only high confidence reactions are retained in the model and medium confidence reactions are included in certain conditions
mCADRE	Transcriptomic data is used to identify core reactions and zero expression reactions. All core reactions are retained in the model unless supported by a number of zero expression reactions
GIMME	Transcriptomic data is used to define low-expressed reactions. The low-expressed reactions are used at minimal level keeping the objective function above a certain level
iMAT	High and low expression reactions are identified using any omics data. An optimal trade-off needs to be obtained between including high expression reactions and excluding low expression reactions
INIT	Weights are defined using any omics data and an optimal trade-off is obtained between including and excluding reactions based on their weights
FASTCORE	Any omics data can be used to define one set of core reactions that are active in the extracted model. It also finds the minimum number of supporting reactions to the core reaction
Metabo tools	Extracellular metabolome, transcriptome and proteome data can be integrated in the model

Thus, we can generate a condition-specific GSMMs reflecting specific properties of a specific cell like a cancer cell or a microbial strain producing a particular metabolite.

8.6 Kinetic Models

These models do not maintain steady state unlike GSMM and allows us to study the system as a dynamic system. It requires knowledge about initial concentration of metabolites and kinetic reaction coefficients. Therefore, this framework can capture the metabolic state of a small system, whereas GSMM reflects the entire metabolic profile of a cell. The overall size of a kinetic model is much smaller than a genome-scale model. Here, the focus is to model few reactions in deeper mechanistic details. For instance, we can develop a kinetic model describing the rate and affinities of multiple reactions. A set of reactants participating in metabolic reactions in a cell is defined by deterministic methods such as non-linear differential equations or partial differential equations. However, some biological processes with few participating molecules such as cell signalling and gene expression are better explained using stochastic kinetic models rather than conventional deterministic models. The deterministic equations describe the alterations in intracellular metabolite concentration over a defined time period. The computational solutions obtained by kinetic modelling helps us in understanding the dynamics of a biological process. However, the biological information obtained is very limited in kinetic models due to small number of reactions (often less than 20) included in the model. These models include various reaction kinetic parameters like enzyme concentration (E_0), turnover number (K_{cat}) and Michaelis constant (K_m), which are measured experimentally using biochemical methods. In case the estimation of kinetic parameters is not possible, computational methods such as parametric estimation methods and Monte Carlo methods are implemented to obtain the value of the kinetic parameters. The kinetic model can be integrated with the genome-scale models forming large kinetic genome-scale models. Overall, a kinetic model often has a large number of parameters. The kinetic models of metabolic reactions are constructed using various computational approaches such as mechanistic modelling, thermokinetic modelling, modular and convenience kinetic rate law modelling, ensemble modelling, optimization and risk analysis of complex living entities (ORACLE), structural kinetic modelling (SKM) and mass action stoichiometric simulation (MASS). Most of these approaches excluding mechanistic and ensemble models can be applied to develop a large-scale genome model. However, both mechanistic and ensemble methods have a potential to build a kinetic model with a maximum of 50 reactions. For example, a whole cell kinetic erythrocyte model of many healthy individuals was developed recently using the MASS approach. This kinetic model is useful in predicting susceptibility to the hemolytic anaemia often induced by an antiviral drug.

8.7 Motifs

Motifs are recurring patterns found in significantly higher frequency in the real network in comparison to an ensemble of randomized networks (Fig. 8.8). Randomized networks have all the characteristic features of a real network such as total number of nodes and edges except the connection between nodes made on a

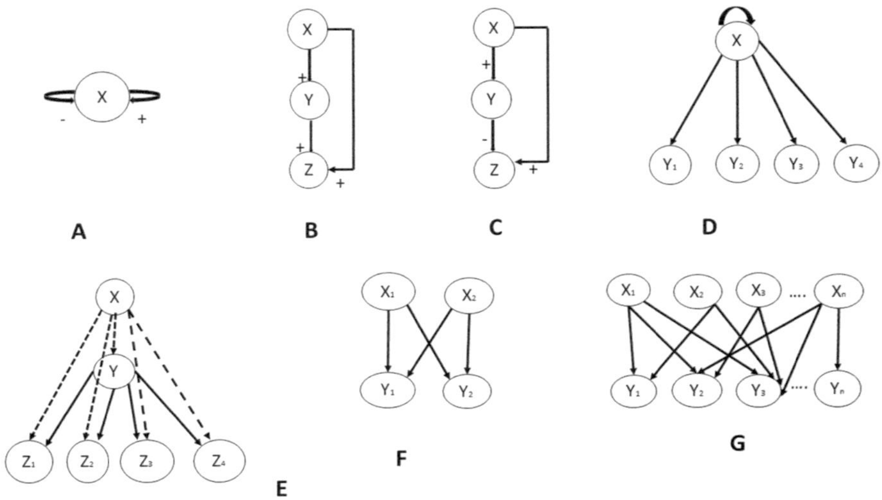

Fig. 8.8 Common network motifs in the biological networks: (**a**) Positive and negative autoregulation (**b**) Coherent feed-forward loop (**c**) Incoherent feed-forward loop (**d**) Single-input module (**e**) Multi-output feed-forward loop (**f**) Bi-fan and (**g**) Dense overlapping regulons (DOR)

random basis. The regulation of a gene by its own product is known as autoregulation and about 59% of transcription factors in *E. coli* are autoregulated. This is the simplest example of motif concerned with a single node. Presence of self-edges on a node is an example of positive and negative autoregulation in a transcriptional network. The negative autoregulation (auto-repression) is more frequent network motif than the positive autoregulation (auto-activation). It not only speeds up the response time of the gene circuits but also provides robustness to fluctuations in the production rate of a gene. On the other hand, positive autoregulation is less common network motif and constitutes about 10% of transcription factor in *E. coli*. The positive autoregulation has an opposite effect to negative autoregulation by slowing down the response time of a gene circuit which may be useful in stage-specific production of gene products during prolonged developmental processes. Moreover, the feed-forward loop (FFL), a very strong motif with three nodes is more abundant in the transcriptional networks of *E. coli*, yeast and higher organisms. A feed-forward loop consists of three transcription factors X, Y and Z. Z is regulated by X directly and is also regulated indirectly by X through Y. The three edges of the FFL may indicate regulation by activation (positive sign) or repression (negative sign). The FFL may be classified into coherent and incoherent types based on the positive and negative signs of the edges. Both direct path and indirect path to Z from X has same signs in the coherent FFL, whereas sign of the direct path from X is opposite to the sign of indirect path from X in incoherent FFL. Most abundant type of FFL is coherent FFL in the biological networks, where all edges are positive followed by incoherent FFL where direct positive regulation is opposed by indirect negative regulation. Coherent FFL usually filters out slight fluctuations in the input signals and thereby stabilizes cellular gene expression under ever-changing stimuli. On the other hand, incoherent FFL helps in the generation of pulse and speeds up the response time. It seems that same FFL is rediscovered by convergent evolution again and again in the biological networks of different organisms due to its vital cellular function.

In addition, there are two larger families of motifs known as single-input module (SIM) and dense overlapping regulons (DOR). In SIM network motif, a single master transcription factor performs dynamical function by controlling several target genes concomitantly. It regulates the temporal expression of genes in a defined order based on their protein product requirement in the cell. In SIM circuit, the gene activated first is deactivated in the last and this temporal order is known as the last-in-first-out (LIFO). This order is observed experimentally in the arginine biosynthesis of *E. coli* where individual genes are expressed at an interval of 5–10 min. However, same activation and deactivation order of genes is desirable in some cases and is known as the first-in-first-out (FIFO) order. The multi-output FFL can generate the temporal FIFO order. For example, the multi-output FFL regulates the expression of motor proteins of flagella in *E. coli*. Bi-fan is a four-node overlapping pattern where two transcription factors X_1 and X_2 control jointly the expression of two target genes Z_1 and Z_2. The dense overlapping regulons (DOR) is a set of input transcription factors regulating through a dense overlapping wiring to another set of target genes. It is a combinatorial decision-making device having multiple input functions in order to regulate a series of target genes. There is a large number of DORs regulating hundreds of genes in the transcriptional networks of *E. coli* and yeast. The DORs are often composed of other common motifs like SIMs and FFLs.

The neuronal networks of *C. elegans* have similar FFLs akin to transcriptional networks although the first one has a much higher spatial scale than the latter one. Another interesting similarity is the presence of multi-input FFLs in neuronal network instead of multi-output FFLs in the transcriptional network.

8.8 Robustness

Robustness is the inherent property of a complex biological system which protects an organism from environmental and genetic perturbations. The traits contributing to the robustness of a system and thereby enhancing overall fitness in an organism are favoured by major evolutionary forces like natural selection and genetic drift. Robustness does not mean staying unchanged under the perturbations. Instead, it allows certain changes in the structure and components of the system in order to maintain the specific function of the system. There are four basic mechanisms involved in the maintenance of robustness in the system: system control, modularity, decoupling and alternative mechanisms. A system is controlled by both positive and negative feedback mechanisms as a robust response to perturbations. The negative feedback mechanism often plays a predominating role in enabling robust responses. For example, bacterial chemotaxis uses negative feedback in response to a variety of external stimuli. Modularity plays an important role in containing perturbations locally in order to minimize its impact on the whole system. For example, cell is a physical module of a multicellular organism. Decoupling separates the low-level variations such as mutations from its high-level expressions such as phenotypes. For example, Hsp90 decouples mutations from the overall phenotype of an organism and not only provides a rich genetic diversity but also buffers against harmful mutations. An alternative or fail-safe mechanism provides several means to accomplish a specific function. This mechanism includes redundancy, diversity and overlapping functions. Redundancy is the presence of several identical or similar interchangeable components in a system. If one component fails under a perturbation, the second one takes over the function to achieve the functional integrity of the system. The presence of glycolysis and oxidative phosphorylation in a cell metabolism is a common example of redundancy. Glycolysis and oxidative phosphorylation are well known for production of ATP under anaerobic and aerobic conditions, respectively. Thus, a yeast cell undergoes drastic changes in the energy metabolism based on the availability of either glucose or ethanol as a nutrient in culture medium.

A highly robust biological system is likely to have a point of fragility which turns out to be the vulnerable point for initiation and progression of a disease in an organism. *Diabetes mellitus* and cancer are two major complex diseases manifested by failure of robustness in a biological system. Human being evolved under several environmental constraints such as starvation, high energy demanding hunting lifestyle and always exposed to ever-changing pathogen through ages. Thus, physiological and biochemical system in human evolved in such a way to develop robustness against these environmental perturbations. However, the drastic changes in the lifestyle of contemporary human such as abundant high energy food and low energy requiring lifestyle have paved the way for fragility of this robust system in form of *Diabetes mellitus*. Similarly, cancer is a highly robust disease against anti-cancer drugs and frequently relapse even after chemotherapy. It co-opts the intrinsic mechanisms of robustness in the body such as genetic heterogeneity of tumour cells and inherent responses against hypoxia. Thus, the mechanistic basis of biological robustness can be better understood using a quantitative measure of the robustness such as network robustness measure (NRM) and response ability measure (RAM). The network of cancer cell exhibited better robustness than the healthy cell in terms of NRM, whereas the response ability measure (RAM) was lower for the cancer cell in respect to healthy cell.

8.9 Exercises

1. Scale-free networks are frequently found in biological world. Create a scale-free network and its adjacency matrix with 25 nodes in the R environment and measure degree, degree distribution, global clustering coefficient (transitivity) and diameter of the network.

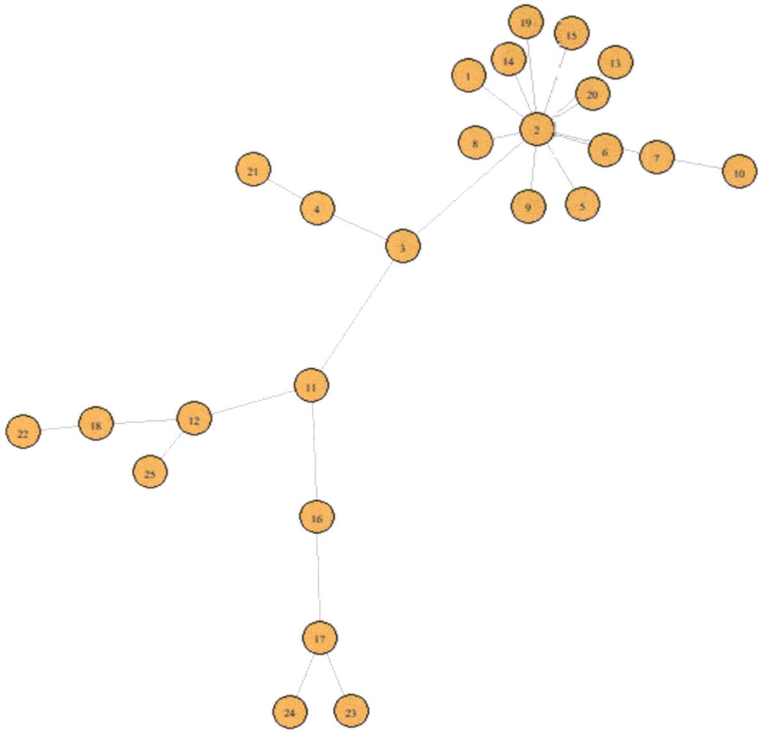

Fig. 8.9 A scale-free network with 25 nodes. Node 2 is the hub node in the network

Solution (Figs. 8.9 and 8.10)

We will load R package igraph and run following scripts for construction of network and analysis.

```
> scale.free <- barabasi.game(25, directed=F,algorithm ="psumtree",
start.graph = NULL)
> plot(scale.free, vertex.label.cex= 0.5, edge.arrow.size=0.5,vertex.
size = 10, layout=layout.fruchterman.reingold)
```

#Fig. 8.9

```
> scale.free.adjacency<-get.adjacency(scale.free)
> scale.free.adjacency
25 x 25 sparse Matrix of class "dgCMatrix"
```

#Fig. 8.10

```
> degree(scale.free)
 [1] 1 12 3 2 1 1 2 1 1 1 3 3 1 1 1 2 3 2 1 1 1 1 1 1 1
> degree_distribution(scale.free)
 [1] 0.00 0.64 0.16 0.16 0.00 0.00 0.00 0.00 0.00 0.00 0.00 0.00 0.04
```

```
 [1,] . 1 . . . . . . . . . . . . . . . . . . . . . . .
 [2,] 1 . 1 . 1 1 1 1 1 . . . 1 1 1 . . . 1 1 . . . . .
 [3,] . 1 . 1 . . . . . . 1 . . . . . . . . . . . . . .
 [4,] . . 1 . . . . . . . . . . . . . . . . . 1 . . . .
 [5,] . 1 . . . . . . . . . . . . . . . . . . . . . . .
 [6,] . 1 . . . . . . . . . . . . . . . . . . . . . . .
 [7,] . 1 . . . . . . . 1 . . . . . . . . . . . . . . .
 [8,] . 1 . . . . . . . . . . . . . . . . . . . . . . .
 [9,] . 1 . . . . . . . . . . . . . . . . . . . . . . .
[10,] . . . . . . 1 . . . . . . . . . . . . . . . . . .
[11,] . . 1 . . . . . . . . 1 . . . 1 . . . . . . . . .
[12,] . . . . . . . . . . 1 . . . . . . 1 . . . . . . 1
[13,] . 1 . . . . . . . . . . . . . . . . . . . . . . .
[14,] . 1 . . . . . . . . . . . . . . . . . . . . . . .
[15,] . 1 . . . . . . . . . . . . . . . . . . . . . . .
[16,] . . . . . . . . . . 1 . . . . . 1 . . . . . . . .
[17,] . . . . . . . . . . . . . . . 1 . . . . . . 1 1 .
[18,] . . . . . . . . . . . 1 . . . . . . . . . 1 . . .
[19,] . 1 . . . . . . . . . . . . . . . . . . . . . . .
[20,] . 1 . . . . . . . . . . . . . . . . . . . . . . .
[21,] . . . 1 . . . . . . . . . . . . . . . . . . . . .
[22,] . . . . . . . . . . . . . . . . . 1 . . . . . . .
[23,] . . . . . . . . . . . . . . . . 1 . . . . . . . .
[24,] . . . . . . . . . . . . . . . . 1 . . . . . . . .
[25,] . . . . . . . . . . . 1 . . . . . . . . . . . . .
```

Fig. 8.10 The adjacency matrix of the network is a binary matrix (0, 1) where each entry in the diagonal is zero. Each dot indicates zero in the matrix

```
#Degree distribution shows the relative frequency of all degrees such as
0,1, 2, 3,4 etc.
> transitivity(scale.free)
[1] 0
> diameter(scale.free)
[1] 7
```

2. An R package sybil is widely used in the constraint-based modelling of metabolic networks. A core metabolic model of *E. coli* is available as a data file with this package. Perform following tasks in the R environment using sybil package:
   (a) Read the basic features of this model, total number of reactions and their reactions IDs using sybil. Plot the stoichiometric matrix of this model in the R environment.
   (b) FBA estimates a single flux distribution in order to solve the optimization problem. Find the total number of exchange reactions with their upper and lower bounds and perform flux balance analysis of the model using the objective function of biomass production.

```
[1]  "ACALD"                    "ACALDt"    "ACKr"       "ACONTa"
[5]  "ACONTb"                   "ACt2r"     "ADK1"       "AKGDH"
[9]  "AKGt2r"                   "ALCD2x"    "ATPM"       "ATPS4r"
[13] "Biomass_Ecoli_core_w_GAM" "CO2t"      "CS"         "CYTBD"
[17] "D_LACt2"                  "ENO"       "ETOHt2r"    "EX_ac(e)"
[21] "EX_acald(e)"              "EX_akg(e)" "EX_co2(e)"  "EX_etoh(e)"
[25] "EX_for(e)"                "EX_fru(e)" "EX_fum(e)"  "EX_glc(e)"
[29] "EX_gln_L(e)"              "EX_glu_L(e)" "EX_h(e)"  "EX_h2o(e)"
[33] "EX_lac_D(e)"              "EX_mal_L(e)" "EX_nh4(e)" "EX_o2(e)"
[37] "EX_pi(e)"                 "EX_pyr(e)" "EX_succ(e)" "FBA"
[41] "FBP"                      "FORt2"     "FORti"      "FRD7"
[45] "FRUpts2"                  "FUM"       "FUMt2_2"    "G6PDH2r"
[49] "GAPD"                     "GLCpts"    "GLNS"       "GLNabc"
[53] "GLUDy"                    "GLUN"      "GLUSy"      "GLUt2r"
[57] "GND"                      "H2Ot"      "ICDHyr"     "ICL"
[61] "LDH_D"                    "MALS"      "MALt2_2"    "MDH"
[65] "ME1"                      "ME2"       "NADH16"     "NADTRHD"
[69] "NH4t"                     "O2t"       "PDH"        "PFK"
[73] "PFL"                      "PGI"       "PGK"        "PGL"
[77] "PGM"                      "PIt2r"     "PPC"        "PPCK"
[81] "PPS"                      "PTAr"      "PYK"        "PYRt2r"
[85] "RPE"                      "RPI"       "SUCCt2_2"   "SUCCt3"
[89] "SUCDi"                    "SUCOAS"    "TALA"       "THD2"
[93] "TKT1"                     "TKT2"      "TPI"
```

Fig. 8.11 A view of all reactants present in the model

(c) In silico gene knockout involves disabling the reactions catalysed by the gene product through setting the upper and lower bound to zero. Find the effects of single-gene knockouts on the objective function of biomass production.

(d) FVA estimates the range of feasible fluxes through each metabolic reaction. Compute and plot flux variability in order to find the flux values for an objective function setting at 80% of maximal biomass production.

Solution (Figs. 8.11, 8.12, 8.13, 8.14, and 8.15)

```
> library(sybil)
Loading required package: Matrix
Loading required package: lattice
> library(glpkAPI)
using GLPK version 4.47
> mp <- system.file(package = "sybil", "extdata")
> model <- readTSVmod(prefix = "Ec_core", fpath = mp, quoteChar = "\"")
reading model description, ... OK
reading metabolite list ... OK
parsing reaction list ... OK
GPR mapping ... OK
```

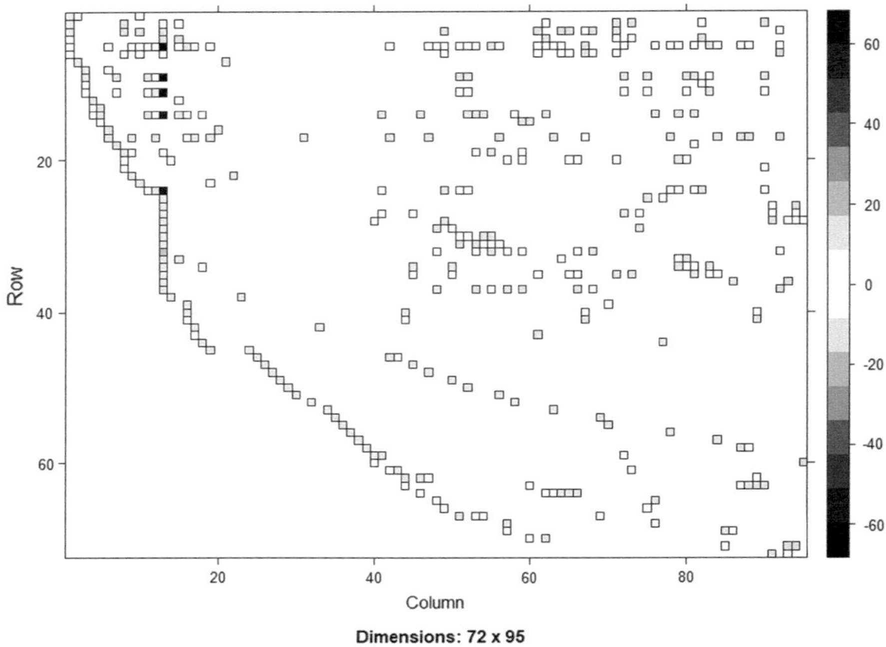

Fig. 8.12 A stoichiometric matrix showing stoichiometric coefficient of each reaction (row) and metabolite (column) pair

```
sub systems ... OK
prepare modelorg object ... OK
validating object ... OK
> model
model name: Ecoli_core_model
number of compartments 2
          C_c
          C_e
number of reactions: 95
number of metabolites: 72
number of unique genes: 137
objective function: +1 Biomass_Ecoli_core_w_GAM
> react_num(model)
[1] 95
> react_id(model)
```

#Fig. 8.11

```
> cg <- gray(0,8/8)
> image(S(model), col.regions = c(cg, rev(cg)))
```

#Fig. 8.12

#	pos.	reaction id	metabolite id	uptake	lb	ub
[1]	20	EX_ac(e)	ac[e]	FALSE	0	Inf
[2]	21	EX_acald(e)	acald[e]	FALSE	0	Inf
[3]	22	EX_akg(e)	akg[e]	FALSE	0	Inf
[4]	23	EX_co2(e)	co2[e]	TRUE		Inf
[5]	24	EX_etoh(e)	etoh[e]	FALSE	0	Inf
[6]	25	EX_for(e)	for[e]	FALSE	0	Inf
[7]	26	EX_fru(e)	fru[e]	FALSE	0	Inf
[8]	27	EX_fum(e)	fum[e]	FALSE	0	Inf
[9]	28	EX_glc(e)	glc_D[e]	TRUE	-10	Inf
[10]	29	EX_gln_L(e)	gln_L[e]	FALSE	0	Inf
[11]	30	EX_glu_L(e)	glu_L[e]	FALSE	0	Inf
[12]	31	EX_h(e)	h[e]	TRUE	-Inf	Inf
[13]	32	EX_h2o(e)	h2o[e]	TRUE	-Inf	Inf
[14]	33	EX_lac_D(e)	lac_D[e]	FALSE	0	Inf
[15]	34	EX_mal_L(e)	mal_L[e]	FALSE	0	Inf
[16]	35	EX_nh4(e)	nh4[e]	TRUE	-Inf	Inf
[17]	36	EX_o2(e)	o2[e]	TRUE	-Inf	Inf
[18]	37	EX_pi(e)	pi[e]	TRUE	-Inf	Inf
[19]	38	EX_pyr(e)	pyr[e]	FALSE	0	Inf
[20]	39	EX_succ(e)	succ[e]	FALSE	0	Inf

Fig. 8.13 Exchange reactions present in the model

```
>findExchReact(model)
```

#Fig. 8.13

```
> opt <- optimizeProb(mod, algorithm = "fba", retOptSol = TRUE)
> opt
solver:                        glpkAPI
method:                        simplex
algorithm:                     fba
number of variables:           95
number of constraints:         72
return value of solver:        solution process was successful
solution status:                    solution is optimal
value of objective function (fba):       0.873922
value of objective function (model):     0.873922
> opt<-oneGeneDel(model)
```

[1]	8.739215e-01	8.739215e-01	2.111407e-01	8.739215e-01	8.739215e-01	8.739215e-01
[7]	8.739215e-01	8.739215e-01	8.739215e-01	7.823511e-01	8.583074e-01	8.583074e-01
[13]	8.739215e-01	8.739215e-01	8.739215e-01	3.742299e-01	3.742299e-01	3.742299e-01
[19]	3.742299e-01	3.742299e-01	3.742299e-01	3.742299e-01	3.742299e-01	8.739215e-01
[25]	-2.185145e-28	8.739215e-01	8.739215e-01	8.739215e-01	8.739215e-01	8.739215e-01
[31]	8.739215e-01	-1.024174e-18	8.739215e-01	8.739215e-01	8.739215e-01	8.739215e-01
[37]	8.739215e-01	8.739215e-01	8.739215e-01	8.739215e-01	8.739215e-01	8.739215e-01
[43]	8.739215e-01	8.739215e-01	8.739215e-01	8.739215e-01	0.000000e+00	0.000000e+00
[49]	8.739215e-01	8.739215e-01	8.739215e-01	8.739215e-01	8.638133e-01	3.087151e-20
[55]	8.739215e-01	8.739215e-01	8.739215e-01	8.739215e-01	8.739215e-01	8.739215e-01
[61]	8.739215e-01	8.739215e-01	8.513885e-01	8.739215e-01	8.739215e-01	8.739215e-01
[67]	8.739215e-01	8.739215e-01	8.739215e-01	8.739215e-01	8.739215e-01	-1.233778e-27
[73]	8.739215e-01	8.739215e-01	8.739215e-01	8.739215e-01	8.739215e-01	8.258193e-01
[79]	8.739215e-01	8.739215e-01	2.116629e-01	2.116629e-01	2.116629e-01	2.116629e-01
[85]	2.116629e-01	2.116629e-01	2.116629e-01	2.116629e-01	2.116629e-01	2.116629e-01
[91]	2.116629e-01	2.116629e-01	2.116629e-01	8.739215e-01	8.739215e-01	8.739215e-01
[97]	8.739215e-01	7.966959e-01	7.966959e-01	8.739215e-01	8.739215e-01	8.739215e-01
[103]	8.739215e-01	8.739215e-01	8.739215e-01	8.739215e-01	8.739215e-01	8.631596e-01
[109]	6.914476e-16	8.638133e-01	8.739215e-01	8.739215e-01	8.739215e-01	8.739215e-01
[115]	8.739215e-01	8.707448e-01	8.739215e-01	8.739215e-01	8.739215e-01	8.739215e-01
[121]	8.739215e-01	8.739215e-01	8.739215e-01	8.739215e-01	8.739215e-01	8.739215e-01
[127]	8.142975e-01	8.142975e-01	8.142975e-01	8.142975e-01	8.583074e-01	8.583074e-01
[133]	8.739215e-01	8.739215e-01	8.739215e-01	8.739215e-01	7.040369e-01	

Fig. 8.14 Objective function of each reaction after single-gene deletion

```
compute affected fluxes ... OK
calculating 137 optimizations ...
> lp_obj(opt)
```

#Fig. 8.14

```
> opt <- fluxVar(mod, percentage = 80, verboseMode = 0)
calculating 190 optimizations ... OK
>summaryOptsol(opt, model)
 Min. 1st Qu.  Median    Mean 3rd Qu.    Max.
 -35.34    0.00    0.00   18.82   10.13 1000.00
> plot(opt)
```

#Fig. 8.15

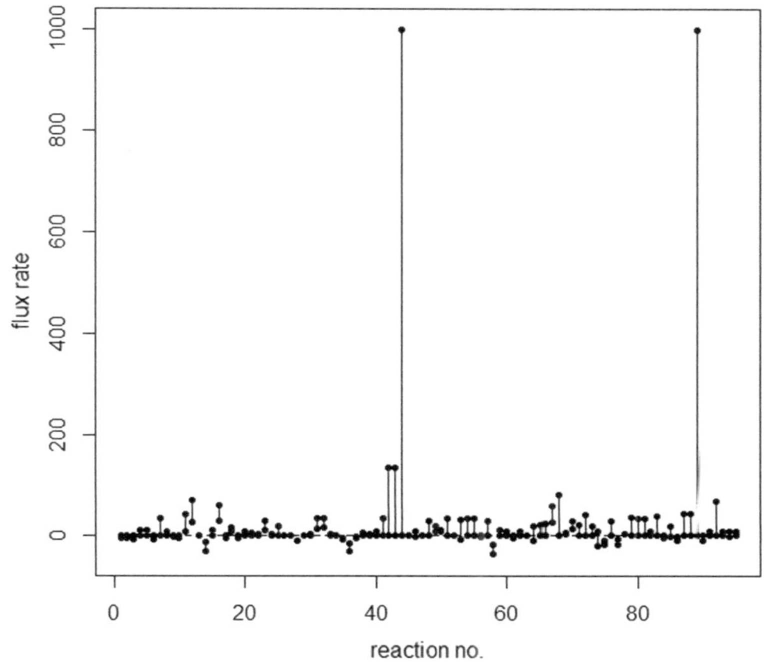

Fig. 8.15 Flux variability analysis showing minimum and maximum flux values of each metabolic reaction

8.10 Multiple Choice Questions

1. A complete graph consists of:
   (a) Multiple nodes without any connection
   (b) Multiple nodes with connection to each other
   (c) Multiple nodes with few connections
   (d) Multiple nodes with some connections
2. Which of the following is/are measure(s) of centrality in a network?
   (a) degree
   (b) eigenvector
   (c) closeness
   (d) All the above
3. The degree distribution of a random network follows:
   (a) Poisson distribution
   (b) Exponential distribution
   (c) Power law distribution
   (d) None of the above

4. The range of values of exponent of a power law distribution in a scale-free network is:
   (a) 1–2
   (b) 2–3
   (c) 3–4
   (d) 4–5
5. Which of the following statement is true regarding objective function?
   (a) Relative contribution of each reaction to the phenotype
   (b) A phenotype of a cell is designated as objective function
   (c) Biomass production or ATP utilization may be an objective function
   (d) All the above
6. The maximum and minimum fluxes through a reaction are measured using:
   (a) Flux mode
   (b) Elementary mode
   (c) Extreme pathway
   (d) Flux variability
7. An abundant motif with three nodes in living organism is:
   (a) Autoregulation.
   (b) Feed-forward loop
   (c) Feed-backward loop
   (d) Single-input module
8. The neuronal network of C. elegans consists of:
   (a) Multi-output FFL
   (b) Multi-input FFL
   (c) LIFO
   (d) FIFO
9. Linear programming during flux balance analysis involves:
   (a) Linear constrained optimization
   (b) Non-linear optimization
   (c) Unconstrained optimization
   (d) Non-linear constrained optimization
10. The optimum expression value of Lac Z protein in *E. coli* cell is:
   (a) 40,000
   (b) 50,000
   (c) 60,000
   (d) 80,000

   Answer: 1. b 2. d 3. a 4. b 5. d 6. d 7. b 8. b 9. a 10. c

Box 8.1 Biochemical Reaction to Algebraic Equation

A metabolic system consists of series of biochemical reactions catalysed by specific enzymes. Thus, computational modelling of these biochemical reactions can only be performed after transforming them into mathematical

(continued)

Box 8.1 (continued)

expressions. For example, we have a pool of metabolite. M with concentration m in a biochemical system. The rate of change in m depends upon the individual rates of reactions involved in either producing or consuming M.

$$V_1 + V_2 + V_3 \rightarrow M \rightarrow V_4 + V_5 + V_6$$

Thus, the rate of change in c at any time point may be written as

$$\mathrm{dm/dt} = V_1 + V_2 + V_3 - V_4 - V_5 - V_6$$

where V_1 is the net rate of the first reaction and V_2 is the rate of the second reaction, etc.

For example, we have another reaction $3X + Y \rightarrow 4Z$.

Alternatively, this reaction may be written as $0 = -3X - Y + 4Z$.where -3, -1, and $+4$ are stoichiometric coefficients. All reactants have negative signs and all products have positive signs.

We can write a general equation using the symbol c for stoichiometric coefficient:

$$0 = c_x X + c_y Y + c_z Z + \ldots.$$

Box 8.2 Stoichiometric Matrix

Stoichiometric matrix is a mathematical expression of metabolic reactions of size m x n. Each row (m) and column (n) of this matrix represents one metabolite and one reaction, respectively. Each element of the matrix is the stoichiometric coefficient of the metabolite participating in a particular reaction. The coefficient is negative for a metabolite consumed in a reaction and positive for a metabolite produced in a reaction. The stoichiometric coefficient is zero if a metabolite is not participating in a reaction. Stoichiometric matrix is usually a sparse matrix because a few metabolites participate in majority of biochemical reactions.

Box 8.3 Linear Programming

Linear programming is a form of constrained optimization where all mathematical expressions are linear. The main elements of a linear constrained optimization process are decision variables, objective function, constraints

(continued)

Box 8.3 (continued)
and variable bounds. The values of the decision variables may not be available initially but can be adjusted to find the best value of the objective function. The objective function is the mathematical expression of a cost function combining all the variables to be treated as an ultimate goal of the optimization process such as biomass or ATP production. The objective function is either minimized or maximized under certain number of constraints. Constraints are mathematical expressions to represent limits in finding the possible solutions. Each variable has an upper and lower bound during the optimization process. When variables have only integer values, solving the linear programming problem is more difficult and is known as integer programming.

Box 8.4 Optimal Design of Gene Circuits
The optimal expression of a protein in a cell is determined by a fitness function that tends to maximize its value. This fitness function in a bacterial population is reflected in the growth rate of cells. An *E. coli* cell produces about 60,000 copies of Lac Z protein which is utilized for breaking down the lactose, a source of energy. These cells utilize this energy to maximize the growth rate. The optimum expression value of Lac Z which is well defined in a particular environment is obtained by the difference between the benefit function and the cost function of producing a protein. Here, benefit is measured in terms of relative increase in the growth rate of the bacteria due to action of Lac Z. On the other hand, the cost function is measured experimentally as the relative decrease in growth rate of bacteria due to expression of Lac Z protein in a lactose-deficient environment induced by a chemical analogue of lactose known as IPTG. High lactose level in the environment allows more production of Lac Z enzyme in *E. coli* than the optimal level of 60,000 per cell because it favours benefit over cost and thus supported by the selection pressure. On the other hand, low level of lactose in the environment reduces the production of enzyme less than the optimal level. Interestingly, the optimal level of enzyme production becomes zero in the absence of lactose in the environment as it incurs only cost without any benefit to the organism. Consequently, the organism will lose the Lac Z gene after few generations in the prolonged absence of lactose in the media.

Box 8.5 LINCS

The Library of Integrated Network-based Cellular Signatures (LINCS) is a NIH-supported systems-level program providing a reference library of signatures related to drugs, genetic perturbation, antibody and disease perturbed human cells. The response to perturbation is measured by various omics technologies such as transcript profiling, cell imaging, mass spectrometry and biochemical methods. The LINCS Data Portal (LDP) is a unified interface to access LINCS Dataset and associated metadata and it contains more than 350 datasets along with 42,000 drugs and other small molecules and 1200 cells. The LDR is a relational database and the central data repository of LINCS datasets and metadata. The main focus of this program is to understand the holistic cellular physiology of various cells and tissues in connection to health and common diseases such as cancer, neurodegenerative disorders and cardiovascular diseases.

Box 8.6 Stochastic Gene Expression

The gene expression is a stochastic phenomenon due to random fluctuations in transcription and translation in spite of a constant environment. These fluctuations may arise due to both intrinsic and extrinsic noises in a biological system. The molecular intrinsic noise is caused by random promoter activation and deactivation, mRNA and protein production and their decay. Although the stochastic gene expression appears to be detrimental to cellular physiology, it provides potential benefits in form of phenotypic cellular diversity. Microbes adapt to a fast changing environment or to a sudden stress due to presence of heterogeneous distinct physiological states without any genetic mutations. This adaptation may play a crucial role in the survival of few microbial cells during antibiotic treatment. DNA methylation and alterations in chromatin structure in eukaryotes can have a profound effect on stochastic transcriptional activity in a particular cell or across a population of cells. The heterochromatin, a dynamic structure, is an inactive chromatin state and also shows variable transcriptional activity.

Box 8.7 Recon3D

Recon3D is the most comprehensive human metabolic network model endowed with three-dimensional structural information of both metabolites and proteins along with large-scale phenotype data. This computational model

(continued)

Box 8.7 (continued)
allows us to characterize various mutations associated with a disease. Moreover, the metabolic response to certain drugs can also be identified using this model. A total of 3288 open reading frames, 13,543 metabolic reactions, 4140 unique metabolites and 12,890 protein structures are included in this model. This computational resource is useful in investigating the molecular basis of human metabolism in health and diseases.

Summary

- Systems biology is an emerging discipline to understand the holistic biology at the systems level.
- Complex biological systems exhibit complex dynamic behaviour at different scales of biological organization.
- A good computational model only captures the fundamental features of a system ignoring other irrelevant details.
- The biological networks with a power-law degree distribution are known as scale-free networks.
- Flux-balance analysis estimates the flow of metabolites in a metabolic network and predicts the growth rate of an organism or rate of production of a useful metabolite.
- Motifs are recurring patterns found in higher frequency in the real network in comparison to an ensemble of randomized networks.
- Coherent FFL is the most abundant type of feed-forward loop in the biological networks.
- Robustness allows certain changes in the structure and components of the system in order to maintain the specific function of the system.

Suggested Reading

Newman MEJ (2010) Networks: An introduction. Oxford University Press, Oxford

Kovert MW (2015) Fundamentals of Systems Biology: From synthetic circuits to whole cell model. CRC Press, Boca Raton

Alon U (2019) An introduction to systems biology: Design principles of biological circuits, 2nd edn. Chapman and Hall/CRC, Boca Raton

Marian Walhout AJ, Vidal M, Dekker J (2013) Handbook of Systems Biology. Academic Press, Cambridge

Clinical Bioinformatics

9

Learning Objectives
You will be able to understand the following after reading this chapter:

- Role of bioinformatics in clinical research.
- Big data in medicine.
- Different approaches of next-generation sequencing to identify genomic variations associated with a disease.
- Computational models of complex diseases.
- Computational discovery of network biomarkers.
- Discovery of multi-target drugs.
- Application of artificial intelligence in medicine.
- Genome-wide association mapping.

9.1 Introduction

The clinical bioinformatics is an integrative multidisciplinary approach to combine natural sciences, computer science and medicine in such a way that datasets are transformed into knowledge useful in clinical practices. Biological systems are largely complex and non-linear in nature. A complex system consists of numerous components such as genes, proteins and metabolites at each scale. For example, genomic scale consists of thousands of genes and proteomic scale has numerous protein molecules as components. These components in each scale usually interact and function in a coordinated manner. Moreover, there is an interaction across different scales in a multi-scale biological system. Thus, no single omics data like genomics, transcriptomics or proteomics can provide a clear understanding of regulatory networks involved in a disease aetiology. The common diseases like

B. K. Tiwary, *Bioinformatics and Computational Biology*,
https://doi.org/10.1007/978-981-16-4241-8_9

cancer, diabetes and mental disorders have not only a genetic basis but also induced by various epigenetic and environmental factors. The symptoms associated with a certain disease is a result of perturbations in the molecular interaction networks in a multi-scale system. Computational models have a property to reflect the overall complexity associated with a disease. A comparative study of a disease model along with a normal healthy model helps us in understanding the mechanistic details of pathogenesis at a deeper level. These models are generally developed in form of biological networks, genome-scale models and kinetic models.

The clinical bioinformatics has emerged as a powerful discipline due to availability of big data. Most advanced technologies such as NGS, MRI and MS produce a flood of big data which needs to processed and analysed computationally on a regular basis. For example, metabolic profiling gives a holistic picture of all metabolites present in a sample. There is a need for integrating all the data from different sources in order to provide a unified understanding of the data. There are four kinds of data types: structured data, unstructured data, semi-structured data and time-series data. Structured data is stored in a relational database or spreadsheet and accounts for almost 20% of an AI project. Majority of the data used in AI is in form of unstructured data without any predefined formatting like text files and images. However, some data is a hybrid of structured and unstructured data known as semi-structured data such as XML and JSON formats and constitutes about 5–10% of data used in AI project. The fourth type of data is a time-series data which is a combination of structured, unstructured and semi-structured data. Big data is a critical part of an AI project. A big data is characterized by three V-features: volume, variety and velocity. The volume of a big data is often in the scale of the terabytes and usually unstructured. Moreover, big data is highly diverse consisting of mainly unstructured data along with structured and semi-structured data. These data are created at an extremely high velocity and processed in the memory instead of disks. Other Vs associated with big data are veracity (accuracy of data), value (data from reliable sources), variability (data changing over time) and visualization (graphical representation of the data).

9.2 NGS Applications in Clinical Research

The detection of genomic mutations or variations is one of the major applications of NGS in clinical research. These genomic mutations are classified into several types according to their sizes: single nucleotide polymorphism (SNP), insertion and deletion (indel), copy number variation (CNV) and structural variation (SV). The SNP is the smallest mutation in the genome involving only one nucleotide. The detection of SNP in the NGS data is a simple process and involves alignment of test reads to the reference genome. If most of the reads show a mismatch at the same locus, it is likely that a SNP is present in the test genome. Indel is either an insertion or deletion or a combination of both in the genome. The size of indel varies from one nucleotide base to 1 kbp. The detection of indel in a genome is a more complicated process. The short reads covering the indel region do not align with the reference genome and therefore labelled as unmapped reads. These unmapped reads are broken into two to three smaller parts which are finally aligned to the reference sequence using Smith–Waterman algorithm in order to identify indels in the test genome. Structural variations are either insertions or deletions having a size more than 100 kbp. There are different kinds of structural variations in the human genome known as deletion, duplication, insertion, inversion and translocation. Methods to detect structural variations are based on paired-end mapping, split read, de novo assembly of genome and read depth. Copy number variation (CNV) is a kind of structural variation involving either a duplication or deletion of a genomic segment more than 1 kbp size. There are many diseases known to be associated with the CNV in human such as cancer, autism, schizophrenia and Alzheimer disease. Methods for CNV detection are largely based on either significant increase or significant decrease in read depth signal. The popular R packages for CNV detection are CNV-seq, readDepth and SeqCBS.

Although whole-genome sequencing is most comprehensive method to identify different kinds of genomic variations, targeted sequencing is the most cost-effective method to detect Mendelian disorders and complex diseases. The first step in targeted sequencing is known as target enrichment which is based either on PCR-amplicons or hybridization capture. Targeted sequencing consists of two forms: whole-exome sequencing and targeted deep sequencing. Targeted deep sequencing is further classified into sequencing with multi-gene panel and designed regions. Whole-exome sequencing (WES) covers only 2% of the genome encoding protein sequences but about 85% of genomic variations associated with diseases are localized in this region. It has successfully identified causal gene for Miller syndrome in a very small population consisting of only four individuals and mutations for autism in a large population consisting of 2500 families. However, regulatory regions of the genome located in introns cannot be explored using WES approach. Targeted deep sequencing requires the prior knowledge of the disease and is a better technology than WES in dealing with regulatory regions. Multi-gene panel sequencing is based on array hybridization and covers large number of candidate genes along with introns. The number of genes varies between 70 and 377 in genes panels for diseases such as muscular dystrophy, cardiomyopathy and epilepsy. Similarly, a desired gene or a region can be captured on custom basis using PCR-amplicons or hybridization. Targeted sequencing is often used as a validation step after WES because an average coverage of 1200-fold can be achieved using this approach. The size of NGS sample varies from small number of cases to a large population. A small size of NGS sample is adequate for a common variant but a large population-based study is required for rare variants of complex diseases. Family-based whole-genome or whole-exome sequencing needs about 200 pedigrees for efficient detection of Mendelian disorders. However, this family-based approach is not suitable for complex diseases. Case-control deep sequencing needs 125 to 500 case-control pairs for high detection power but is not sufficient for a novel gene discovery. However, a case-control whole-genome or whole-exome sequencing using 500–1000 case-control pairs is capable for discovery of novel genes. Most cost-efficient approach is to perform case-only exome sequencing of 100–1000 subjects but identification of new susceptibility loci is difficult using this approach.

9.3 Network Medicine

The network models are very useful in understanding the overall pathobiology of a disease. The initiation and progression of a disease can be explicitly captured in form of network module. Each disease is represented by a well-defined disease module. The genes involved in a disease are likely to be connected in a disease module. This module consists of an ensemble of directly connected genes performing a certain biological function. Any disruption in this disease module may lead to manifestation of disease. The neighbouring directly interacting gene of a disease-causing gene is likely to play some role in the pathogenesis as well. These network models are developed using the known interactions from literature, RNA-Seq, microarray and yeast two-hybrid data. If the candidate genes of a disease are known, their inclusion in the model makes the modelling process more accurate and effective. There are four types of statistical methods employed in reconstruction of a gene network from microarray gene expression data. These are probabilistic network-based methods such as Bayesian networks, correlation-based methods, partial correlation-based methods and information theory-based methods.

We can identify a disease module including the neighbouring genes of a candidate gene of a network. The average degree of disease genes is higher than the average degree of control genes. Moreover, disease genes involved in a certain disease are likely to interact with the disease genes of other diseases. The average shortest path of disease proteins is lower in protein–protein interaction network. A disease gene does not encode a hub protein and is localized in the periphery of the network. However, a static disease network cannot capture the dynamic changes occurring during the progression of a disease. The dynamic rewiring in the molecular interactions during different stages of a disease is compared with the healthy state using differential network analysis (Fig. 9.1). It is expected that different networks

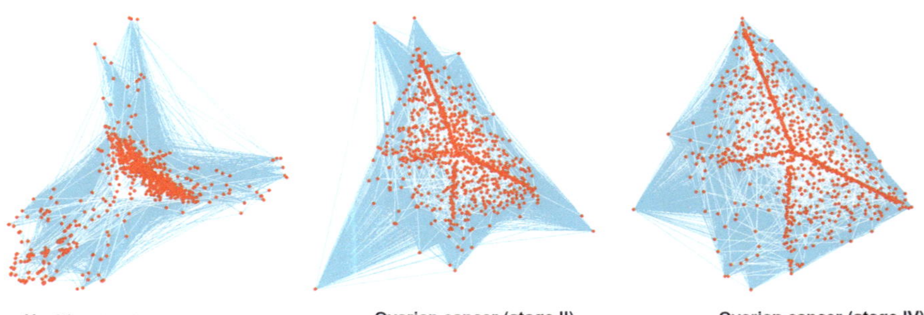

Fig. 9.1 Gene coexpression network of cancer biomarker genes in healthy ovary and ovarian cancer (stage II and IV) showing differential connectivity among genes

representing various disease states show differential connectivity in a timescale barring some highly conserved housekeeping interactions. Even, a particular functional module may be associated with a certain disease and have a potential to be developed as a network biomarker.

Many computational models have been developed for a better understanding of complex diseases such as cancer, type 2 diabetes and various forms of psychiatric diseases. The initiation and progression of various types of cancer is a stepwise process in a timescale and can be modelled as a gene regulatory network from gene expression data from healthy and cancer cells. Gene coexpression analysis has demonstrated that upregulated genes in the lung cancer are highly connected with high degree of centrality. Similarly, the global gene expression in the liver of diabetic patients has established the role of thyroid hormone in alteration in gene expression during type 2 diabetes. Different types of mental disorders such as schizophrenia, bipolar disorder and depression have shown differential connectivity among candidate genes when gene coexpression networks were reconstructed from post-mortem brain. In addition, a genome-scale metabolic model representing a specific type of cancer cell may help in the identification of a drug target for various types of cancer cells. For example, a renal cancer cell having deficiency of fumarate hydratase was modelled in silico and computational models of six hepatocellular carcinoma patients were developed to predict 101 antimetabolites having anti-tumour properties.

9.4 Biomarker Discovery

The development of molecular biomarkers for diagnosis of a disease is one of the major challenges in medicine. Moreover, this task becomes more challenging in case of complex diseases due to overlapping symptoms. Network biomarkers are useful in diagnosis of complex diseases like cancer. EdgeMarker is a computational approach to identify differentially expressed gene pairs known as edge biomarkers. There are successful attempts to develop network biomarkers for diagnosis of early stages of a disease. Dynamical network biomarker (DNB) is an algorithm to detect early warning signals from both single sample and multiple samples before actual onset of a disease. It has been successfully implemented in diagnosis of early stages of lymphoma cancer and liver cancer using high-throughput data. However, it is yet to be applied in a clinical setup with small number of samples. A machine learning method like support vector machine is usually implemented in classification of disease samples from healthy samples using genomic biomarkers. A SVM model can discriminate cancer patients before and after chemotherapy based on the blood levels of plasma nitrite and von Willebrand factor.

9.5 Multi-Target Drugs

Computational modelling is playing a major role in the identification of novel drug targets. It is well known that cancer cells develop resistance to multiple drugs after treatment. Drugs are also having multifarious effect in a cellular environment. Thus, network model provides an effective framework to understand not only the effect of a drug on the target but also side effects associated with the drug. A new paradigm has emerged recently in drug discovery process known as polypharmacology. It emphasizes on the development of network models including multiple drugs and their interaction with multiple targets. A bipartite network shows direct connection between a drug network and its target protein network. Interestingly, high degree nodes in the drug network are connected with high degree nodes in the target protein network. Even the drugs acting on same target protein or direct neighbours in the network model share their side effects. Moreover, the side effects of a drug are caused by corresponding target protein with high degree and betweenness. Cancer cells always find an alternative pathway of proliferation when exposed to a single drug. Therefore, drug combination therapy is an essential approach in cancer treatment. But it is not feasible to test a combination of more than 200 cancer drugs. Combenefit is a computational method to circumvent this challenge and computes a synergy score between a pair of drugs. Majority of drugs act in coherent manner on the target protein and have synergistic effect. However, few drugs are known to function in an incoherent fashion, simultaneously having both positive and negative effect on the target protein. The positive synergistic coefficient measures the pharmacological effect of a combination of drugs, whereas positive side effect coefficient is a measure of side effects of a combination of drugs.

9.6 Artificial Intelligence in Medicine

Artificial intelligence (AI) is a buzz word today in biomedical research including health diagnostics. It has two components: machine learning and deep learning. The rapid development of AI is attributable to better computational infrastructure, availability of big data, and advent of high-speed GPU in last two decades.

9.6.1 Machine Learning

Machine learning was first developed by Arthur L. Samuel in 1959 and its ultimate goal is to develop a predictive model based on training the dataset using one or more algorithms. Here, we train the model rather than explicitly program a computational task. First, the order of the data is randomized and appropriate algorithm is selected by trial and error. The training data which constitutes about 70% of the complete dataset is used to create relationships in the algorithm. Training phase is followed by the evaluation of the model's accuracy using remaining 30% of the dataset. The parameters in the algorithm are adjusted to fine-tune the model for improving the results. Some hyperparameters which cannot be learned directly from the training process are also adjusted during fine tuning. The algorithms used are usually complex but there is no need to compute these algorithms by the practitioner as plenty of programs already available in the R and the Python. Although there are hundreds of machine learning algorithms available, they can be categorized into four major classes which are as follows:

9.6.1.1 Supervised Learning

The labelled data are generally used in large amount in order to get more accurate results. The common types of supervised learning are classification and regression. Classification divides the datasets based on common features. For example, support vector machine and Naïve Bayes classifier are two methods for classification. Regression methods such as linear regression, decision trees and ensemble modelling find a continuous pattern in the dataset. Linear regression finds relationship between some variables to predict the outcome based on inputs of quality data. Decision trees are suitable for nonnumerical data and make decision based on a probability. Ensemble modelling uses more than one model for prediction for more accurate results.

9.6.1.2 Unsupervised Learning

Unlabelled data are used in unsupervised learning to find a common pattern. For example, clustering like k-Means clustering takes unsupervised data to put similar items in one group. It is suitable method for large datasets. Closely related data are identified based on some metrics like Euclidian distance, cosine similarity and Manhattan distance.

9.6.1.3 Reinforcement Learning

The learning process is improved by trial and error using positive and negative reinforcement. It is widely used in games and robotics.

9.6.1.4 Semi-Supervised Learning

This hybrid approach uses both supervised and unsupervised learning. When small amount of unlabelled data are available, unsupervised data can be labelled using deep learning and converted to supervised data. This process is known as pseudo-labelling. MRI data is usually processed using semi-supervised learning.

9.6.1.5 Support Vector Machines (SVM)

Support vector machine is a supervised kernel-based classification method that has been extensively used to develop biomarkers to discriminate disease samples from healthy samples. This algorithm finds a decision boundary between datasets belonging to two different groups. These models can map both healthy and disease data points in a space and then classify them into two groups based on a hyperplane (Fig. 9.2). Only the hyperplane separating the datasets is learned in the SVM. A hyperplane is a flat one-dimensional line in two dimensions. On the other hand, it is a flat two-dimensional plane in three dimensions. But, we cannot visualize a

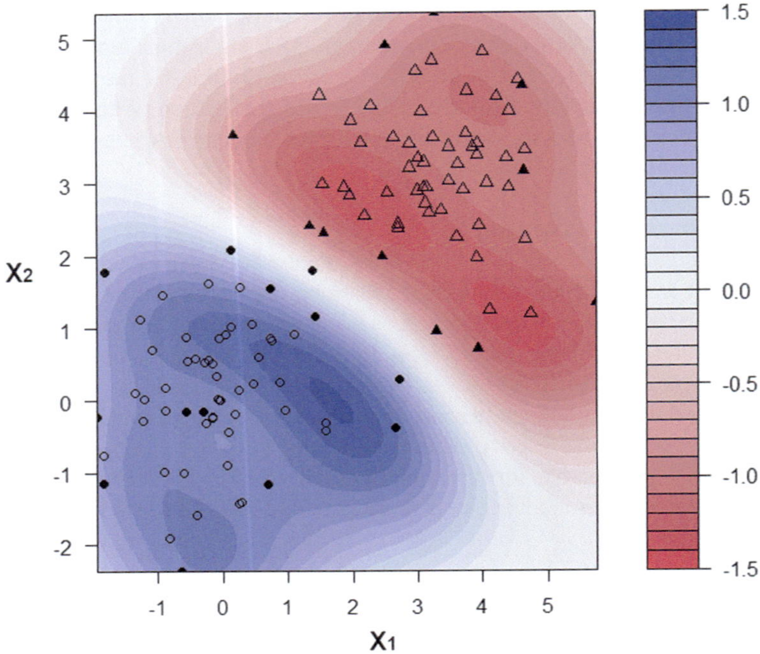

Fig. 9.2 A SVM model classifying healthy (circles) and disease (triangles) samples using radial basis function (RBF) kernel. The maximum margin hyperplane divides both samples and solid data points close to the hyperplane are support vectors

hyperplane in more than three dimensions. The optimal separating hyperplane always maintains extreme minimum distance from the training dataset. A test dataset is evaluated based on its situation on both sides of a maximal margin hyperplane. Any data situated on the incorrect side of hyperplane is treated as misclassification by the algorithm. Thus, the classification is optimized based on the support vectors lying on margin of the hyperplane or on the incorrect side of the margin of their group. The tuning parameter C, the cost of violation to hyperplane margin, is taken for the optimization process using cross-validation. If the tuning parameter C is very small, many data points may violate the hyperplane margin and in turn create numerous support vectors, whereas a large C explores a narrow margin of the hyperplane which is not prone to any violation of hyperplane. In case two groups (i.e. healthy and disease samples) cannot be separated using a linear hyperplane, a polynomial or radial kernel is a best choice to map the two-dimensional data into a higher dimensional space. Support vector machines (SVM) have been widely used for the development of molecular biomarkers as health diagnostics. For example, a support vector machine model has successfully discriminated cancer patients before and after chemotherapy using two blood biomarkers, namely plasma nitrite and von Willebrand factor. However, it is difficult to apply SVM on large datasets and is not suitable for perceptual problems which needs feature engineering like an image classification.

9.6.2 Deep Learning

Deep learning is a subfield of machine learning which analyses the datasets using neural networks mimicking the human brain. The word "deep" refers to the inherent hidden layers in the neural networks directly related to the learning power of the algorithm. Even a single hidden layer neural network is not treated as deep due to presence of a single hidden layer. Here, the goal of learning is to predict a response or classify a response based on certain attributes. An Artificial Neural Networks (ANN) is a function consisting of units called neurons (also known as perceptrons or nodes). Typical feed-forward neural networks have three layers, namely an input layer, a hidden layer and an output layer (Fig. 9.3). The number of neurons in an input layer corresponds to the number of features or variables to be fed in the networks. Similarly, the number of neurons in the output layer corresponds to the number of items predicted or classified. The hidden layer neurons perform non-linear transformation of the input attributes. The relative importance of each neuron in a network is indicated by its value and weight. Each neuron has an activation function and a threshold value to activate the neuron. Thus, a neuron performs summation of weights of all inputs and subsequent activation function in order to pass the output to the next layer. The net input is calculated by the formula which is as follows:

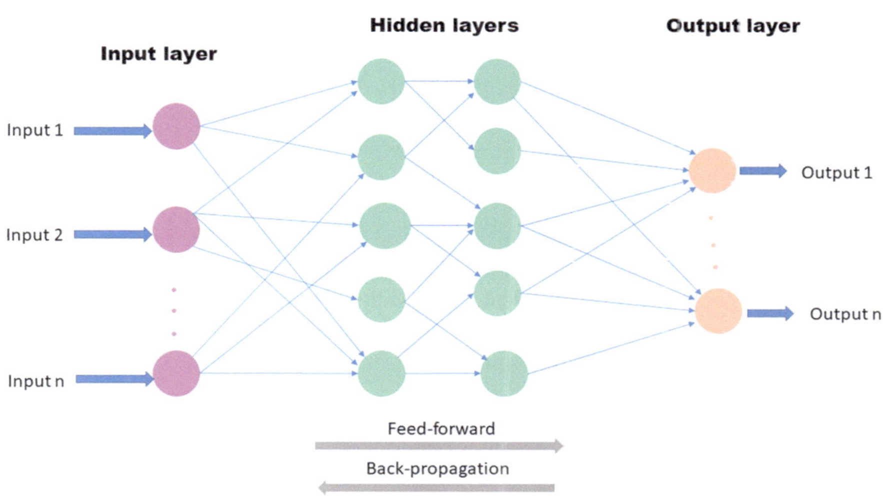

Fig. 9.3 The architecture of an artificial neural network (ANN) consisting of input layer, hidden layers and output layer

$$y_{\text{in}} = x_1.w_1 + x_2.w_2 + x_3.w_3 + x_4.w_4 \ldots\ldots x_m.w_m$$

The value and weight of each neuron is passed to a hidden layer of neurons which uses a function to produce a final output. The activation function is a non-linear function which varies between -1 and + 1 considering the non-linear nature of real world. Each neuron in the networks usually has same activation function. So, the output is computed by applying the activation function over the net input which is as follows:

$$Y = F\ (y_{\text{in}})$$

Most common function is the sigmoid one which produces an output value between 0 and 1. The highest accuracy achieved is 1 for a perfect model. Bias is another constant value used in the calculation of the function and is similar to an intercept in linear regression. It facilitates the activation function of the networks to move either upwards or downwards. This type of training is known as feed-forward neural network. However, this model is too simple and can be improved further by multiple hidden layers resulting in multilayered perceptron (MLP). MLP is endowed with a unique property of backpropagation. The *loss function* or *objective function* compares the output prediction and the true target using a distance score. This score gives a feedback signal to adjust the value of weights to reduce this loss score. The adjustment is done by an optimizer using backpropagation algorithm. Backpropagation involves adjustment of weights in a neural network after getting the errors followed by iterating the new values to optimize the model. For example, one of the input value has an output value of 0.7 indicating an error value of 0.3 (1–0.7). After backpropagation, the output value may improve to 0.75. Thus, the training will continue till output value reaches close to 1. Initially, errors are large due to large weights of the input nodes. After few iterations, the error gradually decreases to find an optimum mid-point value at the bottom of the curve. Most common types of neural networks are recurrent neural networks (RNN), convolutional neural network (CNN) and generative adversarial network (GAN). The recurrent neural networks have a function which processes the input and prior inputs across time as well. The convolutional neural network analyses a complex data like an image data section by section. In GAN, two neural networks compete with each other using a feedback loop and create new object.

The scope of deep learning applications is still narrow as it requires large amount of input data and high computational power. It is a challenging task to select the number of hidden layers and hyperparameters to develop a right model. Usually, comparatively simple machine learning models are found to be more effective using less amount of input data than the complex deep learning models.

9.7 Genome-Wide Association Mapping

Humans have only 0.1% variation of nucleotide bases in their genome. A single nucleotide varies between two individuals in a population and this variation is known as single-nucleotide polymorphism (SNP). The presence of one SNP in a gene can lead to a monogenic disease like sickle cell anaemia. Moreover, some complex diseases such as diabetes and cardiovascular disease are caused by the epistatic interaction among multiple genes and the interaction between the genes and the environment. However, there are two major challenges in understanding the genetic basis of these complex diseases. The discovery of large number of genetic variants along with their modified phenotypic manifestations under different environmental conditions is the first challenge. The second challenge is to define a correct phenotype for a study and proper statistical analysis of the data. There are ten million SNPs estimated in human population. Current commercially available SNP arrays provide a comprehensive genomic coverage of about 660,000 SNPs in different populations. The study design of GWAS involves cases having the disease and controls without the disease and other traits contributing to confounding effect. The SNPs and samples with low quality are usually detected through low call rates and subsequently removed from analysis. In addition, the SNPs with low minor allele frequency and the SNPs deviating from Hardy–Weinberg equilibrium are removed from study in order to avoid false positive findings. It is expected that both allele and genotype frequencies remain stable in a particular population as per requirement of Hardy–Weinberg equilibrium. However, it is advisable to first analyse all SNPs and then examine the validity of Hardy–Weinberg equilibrium for those SNPs associated with the phenotype. A single SNP analysis is conducted using a statistical test to compare the null hypothesis (i.e. there is no association between the SNPs and the phenotype) with the alternative hypothesis (i.e. there is an association between the SNPs and the phenotype). A small p-value obtained during the test leads to the rejection of the null hypothesis and indicates that there is a significant association between the SNP and the phenotype. The genetic effect of the SNP is modelled on a continuous phenotype trait using a linear regression model. The result of the statistical analysis can be visualized using a Manhattan plot highlighting the genomic regions showing $\log_{10}$ (p-value) of the association (Fig. 9.4). Principle components analysis (PCA) is usually used in summarizing the genome-wide variability of SNPs creating principal components of all SNPs in the genome. The first principal component generally captures the population substructure based on ethnicity. There are many programs publicly available for genome-wide association mapping. Some R packages are available on bioconductor for GWAS analysis and analysis of structural variations (Tables 9.1 and 9.2). PLINK is another popular program for whole genome-wide association mapping including various basic analysis such as haplotype analysis and LD estimation. The GWAS Central is a comprehensive database for comparing genotype and phenotype from various genome-wide association studies and appropriate data sets can be retrieved from this database for analysis.

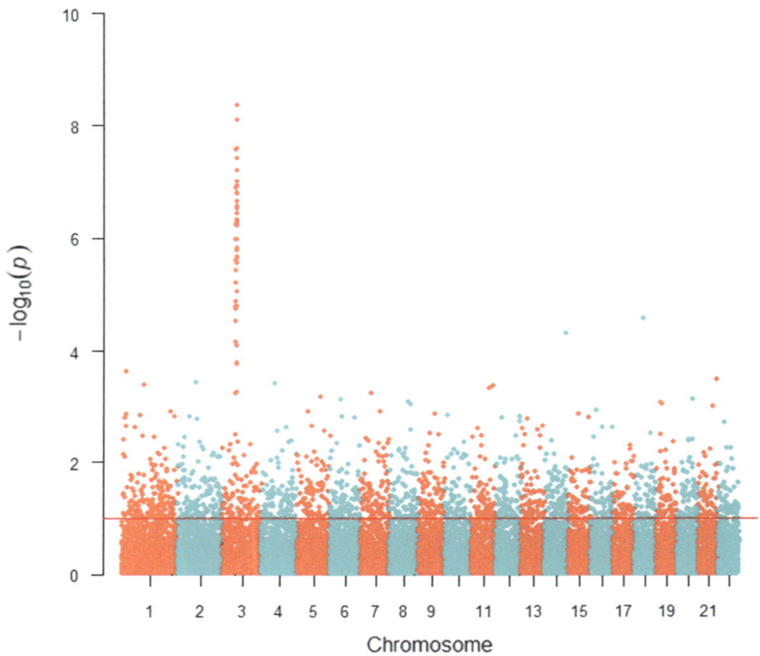

Fig. 9.4 Manhattan plot of a genome-wide association analysis of a disease. X-axis and Y-axis show chromosomal positions and -log10 p values, respectively. The horizontal red line indicates the significance threshold of genome-wide association

Table 9.1 R packages for GWAS analysis

R package	Description
GWASTools	A popular package for quality control and analysis of genome-wide association studies of extremely large datasets.
StatgenGWAS	It performs single trait genome-wide association studies.
IntAssocPlot	It provides integrated visual display of GWAS results with linkage disequilibrium matrix and gene structure.
fastJT	It is useful in feature selection from high-dimensional data in machine learning and genome-wide association studies.
GWAF	It is designed for genome-wide association studies with family data.
bGWAS	It is designed to improve genome-wide association studies using Bayesian approach based on informative priors.
R/fGWAS	A package for functional analysis of genome-wide association studies using Bayesian lasso model.
RAINBOW	It is designed for haplotype-based GWAS without any prior haplotype information.
BGData	It is suite of four R packages, namely BEDMatrix, LinkedMatrix, symDMatrix and BGData for big genomic data sets.

Table 9.2 R packages for analysis of structural variations

R package	Description
Intasv	A package for integrative analysis, annotation and visualization of structural variations
RSVSim	A package for simulation of structural variations
StructuralVariantAnnotator	A package for annotation of structural variants

9.8 Exercises

1. The genotypes of a SNP in a sample are AA, AT, TT, AA, AA, AA, AT, AA, AT, AA, TT, AT. Using the R package *genetics*, compute the allele frequency, genotype frequency and heterozygosity of the genotypes. Plot the frequency of three types of genotypes. Test whether these genotypes are in Hardy–Weinberg equilibrium.

Solution First, the data is converted to genotypes and frequency is computed using the function "summary ()"

```
>library(genetics)
>data<-c("A/A","A/T","T/T","A/A","A/A","A/A","A/T","A/A","A/
T","A/A", "T/T", "A/T" )
> geno <- genotype(data)
> summary(geno)
Number of samples typed: 12 (100%)
Allele Frequency: (2 alleles)
 Count Proportion
A 16 0.67
T 8 0.33
Genotype Frequency:
 Count Proportion
A/A 6 0.50
A/T 4 0.33
T/T 2 0.17
Heterozygosity (Hu) = 0.4637681
Poly. Inf. Content = 0.345679
>plot(geno)
```

```
  > HWE.test (geno)
 ---------------------------------------------
 Test for Hardy-Weinberg-Equilibrium
 ---------------------------------------------
Call:
```

```
HWE.test.genotype(x = geno)
Raw Disequlibrium for each allele pair (D)
 A T
 A -0.05555556
 T -0.05555556
Scaled Disequlibrium for each allele pair (D')
 A T
 A -0.5
 T -0.5
Correlation coefficient for each allele pair (r)
 A T
 A 0.25
 T 0.25
Observed vs Expected Allele Frequencies
 Obs Exp Obs-Exp
A/A 0.5000000 0.4444444 0.05555556
T/A 0.1666667 0.2222222 -0.05555556
A/T 0.1666667 0.2222222 -0.05555556
T/T 0.1666667 0.1111111 0.05555556
Overall Values
 Value
 D -0.05555556
 D' -0.50000000
 r 0.25000000
Confidence intervals computed via bootstrap using 1000 samples
 Observed 95% CI NA's Contains Zero?
 Overall D -5.555556e-02 (-1.875000e-01, 8.333333e-02) 0 YES
 Overall D' -5.000000e-01 (-2.840000e+00, 3.333333e-01) 1 YES
 Overall r 2.500000e-01 (-3.333333e-01, 8.222222e-01) 1 YES
 Overall R^2 6.250000e-02 ( 7.061648e-05, 6.760494e-01) 1 *NO*
Significance Test:
 Exact Test for Hardy-Weinberg Equilibrium
data: geno
N11 = 6, N12 = 4, N22 = 2, N1 = 16, N2 = 8, p-value = 0.5176
```

The p-value indicates that genotypes are in Hardy–Weinberg equilibrium.

2. The linkage disequilibrium (LD) is non-random association among different alleles at different loci in a population. There are two measures of LD. The D′ is the normalized coefficient of linkage disequilibrium (D), whereas R^2 is the measure of correlation coefficient between two loci.

Using the example file in the R package snpStats, compute linkage disequilibrium (LD) among the SNPs in European and African populations based on D′ and R^2

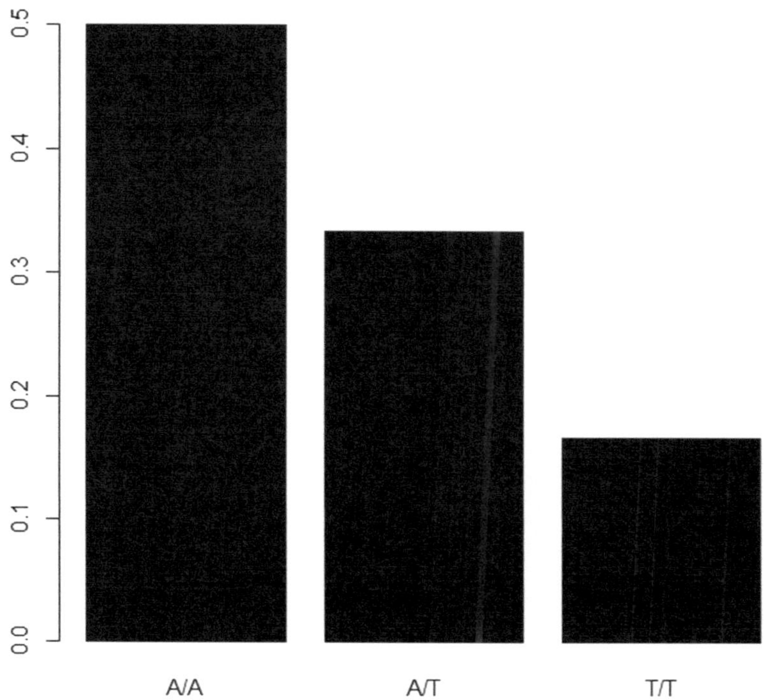

Fig. 9.5 Plot of three types of genotypes

measures. Compare the LD map in both populations based on D′ plus R^2 as well as on R^2 alone (Figs. 9.5, 9.6, 9.7, 9.8, and 9.9).

```
> library(snpStats)
Loading required package: survival
Loading required package: Matrix
> data(ld.example)
> ceph.1mb
A SnpMatrix with 90 rows and 603 columns
Row names: NA06985 ... NA12892
Col names: rs5993821 ... rs5747302
> head(support.ld)
 dbSNPalleles Assignment Chromosome Position Strand
rs5993821 G/T G/T chr22 15516658 +
rs5993848 C/G C/G chr22 15529033 +
rs361944 C/G C/G chr22 15544372 +
rs361995 C/T C/T chr22 15544478 +
rs361799 C/T C/T chr22 15544773 +
rs361973 A/G A/G chr22 15549522 +
```

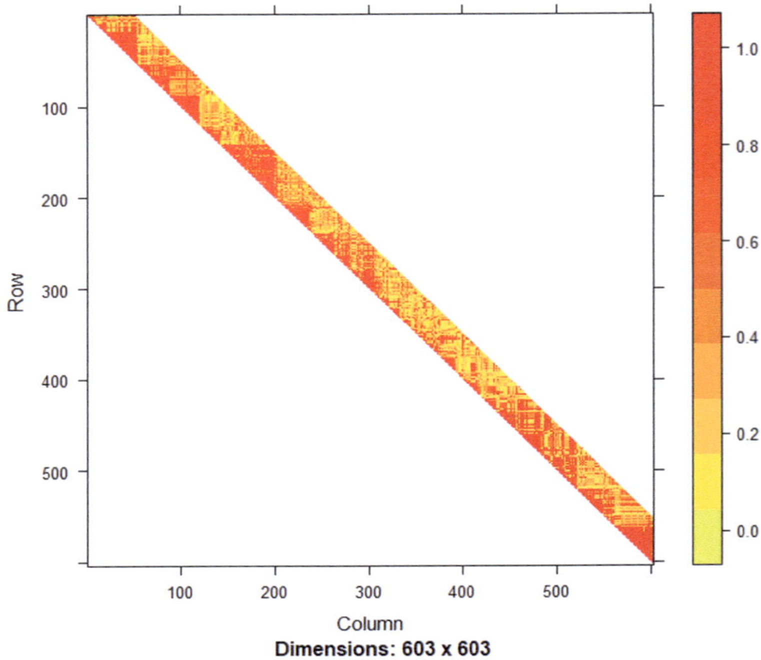

Fig. 9.6 Linkage disequilibrium in European population based on D′ and R^2 measures

```
> ld.afro <- ld(yri.1mb, stats=c("D.prime", "R.squared"),
depth=50)
> ld.euro <- ld(ceph.1mb, stats=c("D.prime", "R.squared"),
depth=50)
> spectrum <- rainbow(10, start=0, end=1/6)[10:1]
> image(ld.euro$D.prime, lwd=0, cuts=9, col.regions=spectrum,
colorkey=TRUE)

> image(ld.afro$D.prime, lwd=0, cuts=9, col.regions=spectrum,
colorkey=TRUE)

> use <- 50:300
> image(ld.euro$R.squared[use,use], lwd=0)

> image(ld.afro$R.squared[use,use], lwd=0)
```

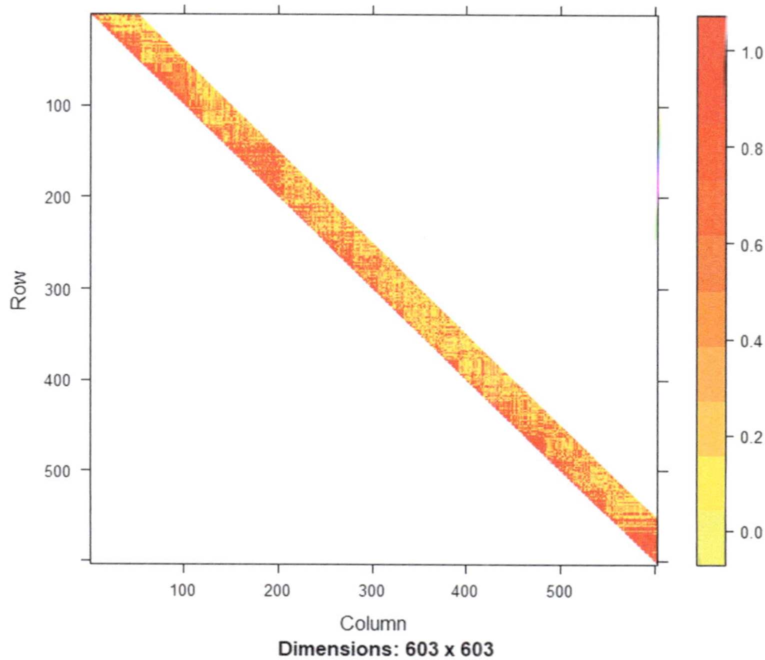

Fig. 9.7 Linkage disequilibrium in African population based on D′ and R^2 measures

Linkage disequilibrium is more noticeable in the European population than in the African population in all comparisons.

9.9 Multiple Choice Questions

1. A big data is characterized by:
   (a) Volume
   (b) Variety
   (c) Velocity
   (d) All the above
2. Copy number variation (CNV) is a structural genomic variation involving more than:
   (a) 1 kbp
   (b) 10 kbp
   (c) 20 kbp
   (d) 25 kbp
3. Whole-exome sequencing for discovery of a novel gene requires:
   (a) 100–200 case-control pairs
   (b) 300–400 case-control pairs

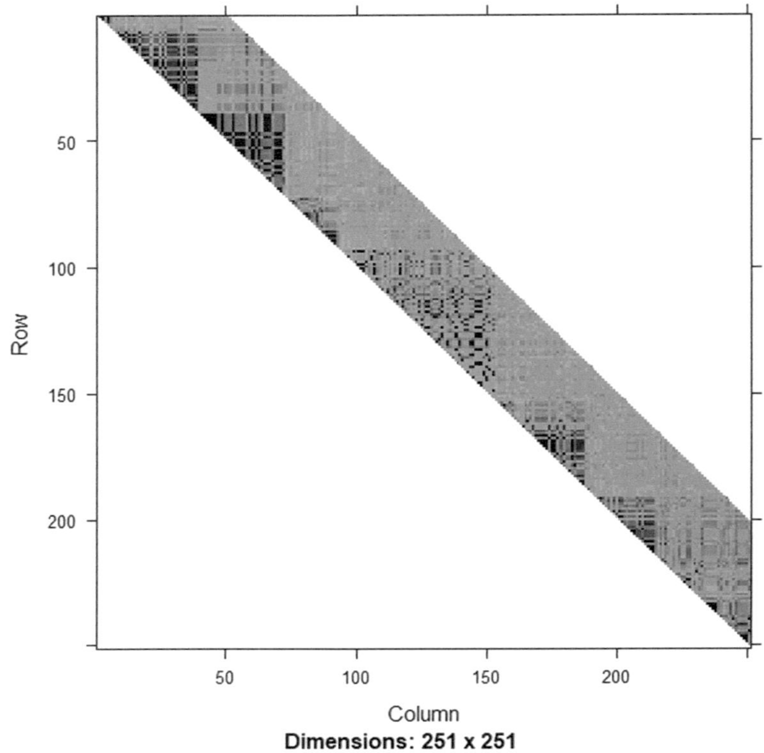

Fig. 9.8 Linkage disequilibrium in European population based on R^2 measure

(c) 500–1000 case-control pairs
(d) 2000–3000 case-control pairs

4. The machine learning algorithm not based on supervised learning is:
   (a) Support vector machine
   (b) Naïve Bayes classifier
   (c) k-mean clustering
   (d) Linear regression
5. The activation function of artificial neural network is:
   (a) Linear
   (b) Non-linear
   (c) Both linear and non-linear
   (d) None of the above
6. The backpropagation is the property of:
   (a) Multilayered perceptron
   (b) Feed-forward neural network
   (c) Feed-backward neural network
   (d) None of the above

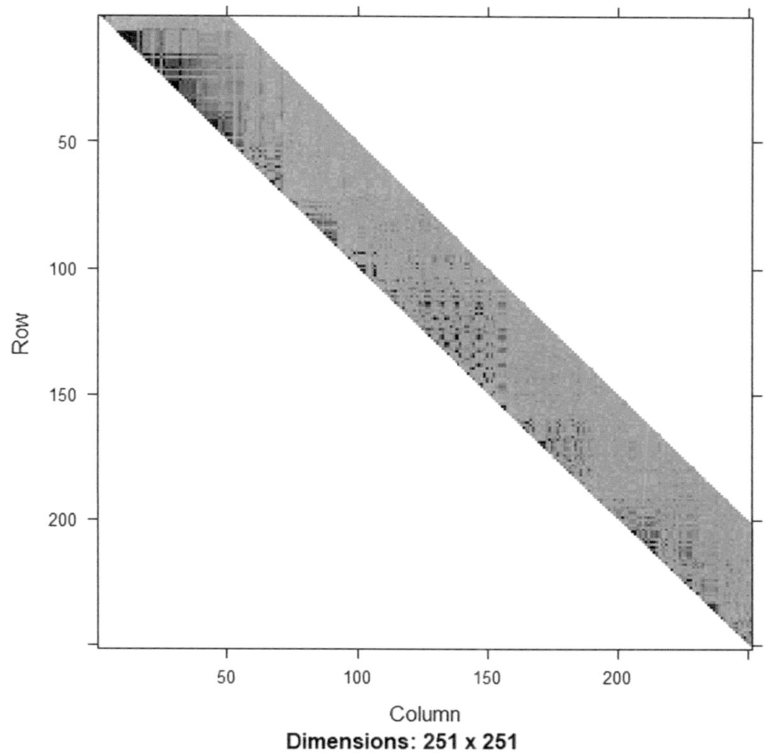

Fig. 9.9 Linkage disequilibrium in African population based on R^2 measure

7. The total number of SNPs estimated in human population is:
   (a) one million
   (b) five million
   (c) ten million
   (d) 15 million.
8. All SNPs associated with the phenotype are expected to follow:
   (a) Hardy–Weinberg equilibrium
   (b) Hardy–Weinberg disequilibrium
   (c) Selection and migration
   (d) Random genetic drift
9. The genome-wide association can be visualized through:
   (a) Scatter plot
   (b) Manhattan plot
   (c) Euclidian plot
   (d) Hamilton plot

10. Which of the following is/are a type (types) of neural network?
    (a) Recurrent neural networks (RNN)
    (b) Convolutional neural network (CNN)
    (c) Generative adversarial network (GAN)
    (d) All the above

Answer: 1. d 2. a 3. c 4. c 5. b 6. a 7. c 8. a 9. b 10. d

Summary

- A comparative study of a disease model and a normal healthy model provides the mechanistic details of pathogenesis at a deeper level.
- A big data is characterized by three V-features: volume, variety and velocity.
- The human diseases associated with the copy number variation (CNV) are cancer, autism, schizophrenia and Alzheimer disease.
- Whole-exome sequencing (WES) covers only 2% of the human genome encoding protein sequences and detects 85% of the genomic variations associated with various diseases.
- The initiation and progression of a disease can be explicitly captured in form of a computational network module.
- A disease gene does not encode a hub protein and is localized in the periphery of the network.
- A genome-scale metabolic model representing a specific-type of cancer cell may help in the identification of a drug target.
- A bipartite network shows direct connection between a drug network and its target protein network.
- Support vector machine (SVM) is a machine learning classification method for development of molecular biomarkers to discriminate disease samples from healthy samples.
- A single SNP analysis compares the null hypothesis (i.e. there is no association between the SNPs and the phenotype) with the alternative hypothesis (i.e. there is an association between the SNPs and the phenotype).

Suggested Readings

Wang X, Baumgartner C, Shields DC, Deng H-W, Beckmann JS (2016) Application of clinical bioinformatics. Springer, Dordrecht

Trent RJA (2014) Clinical bioinformatics. Humana Press, Totowa

Liang K-H (2013) Bioinformatics for biomedical science and clinical applications. Woodhead Publishing, Oxford

Raza K, Dey N (2021) Translational bioinformatics applications in healthcare. CRC Press, Boca Raton

Agricultural Bioinformatics 10

Learning Objectives

You will be able to understand the following after reading this chapter:

- Role of bioinformatics in agriculture.
- Pan-genomic view of various crop species.
- Genome assembly of crop plants.
- Identification of homeologs in crop species.
- Principles and methods of genomic selection for improvement of crops.
- Crop phenomics.
- Computational modelling of agricultural systems.

10.1 Introduction

The estimated population of this world would be about nine billion by the year 2050 and consequently there is a need for a very steep increase in the food production to feed the projected population. So, there is a necessity for large-scale innovations in agriculture to boost its productivity and sustainability as well. Crops have a vital role in global economy being a major source of nutrients for the ever increasing world population. In fact, crop genomics is expected to play a significant role in the second green revolution. The available genomic databases useful in crop genomics research are listed in Table 10.1. The agricultural system is a complex system where the biology of a crop interacts with the environment and management practices. Bioinformatics can play a pivotal role in enhancing the agricultural productivity due to rapid progress in both computing power and next-generation sequencing technology. The interaction and behaviour of the overall agricultural system is better modelled through integrating the data with computational analysis. We can capture

B. K. Tiwary, *Bioinformatics and Computational Biology*,
https://doi.org/10.1007/978-981-16-4241-8_10

Table 10.1 Genomic databases of crops

Database	Description	URL
Gramene	A knowledgebase of genomic and pathway data of major crops	www.gramene.org
LegumeIP	An integrative comparative genomics and transcriptomics database of legume crops	plantgrn.noble.org/LegumeIP
SoyBase	A database of soybean genetic and genomic data	soybase.org
GreenPhyID	A database of comparative pan-genomics of plants including major crops	www.greenphyl.org

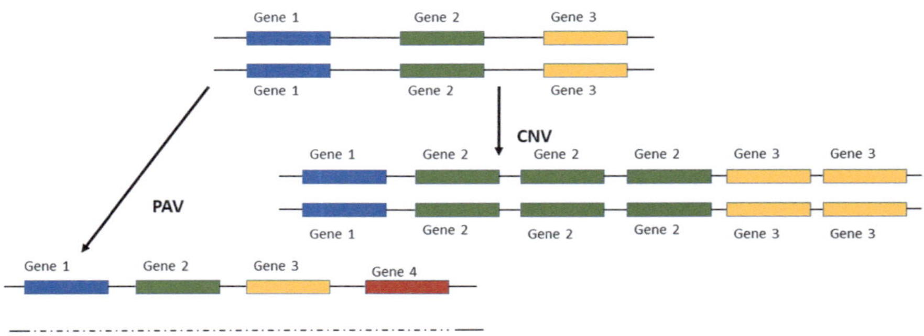

Fig. 10.1 Structural variants like CNV and PAV cause variations in the functional gene content of crop plants. Here, CNV in the genome is represented by three copies of Gene 2 and two copies of Gene 3. The PAV indicates loss of one copy of each gene with gain of another gene, Gene 4 in the genome

a wide variety of real-time big data of different genomic and environmental variables using existing sensor technologies. Some of the promising areas of agriculture having significant attention of bioinformatics research are pan-genome of crops, assembly of plant genomes, identification of homeologs, genomic selection and modelling of agricultural systems.

10.2 Pan-Genome of Crops

Modern crops are endowed with a wide variety of small to large continuum of genomic variations in the form of single-nucleotide polymorphisms (SNPs)-insertions and deletions (indels)-structural variants (SV). SNPs have a major effect on the functional gene content of a crop genome either through premature stop codons or changes in the key functional sites. Similarly, small indels can have a major impact on the genome variation caused by frameshift changes and premature stop codons in the genes. The structural variants such as copy number variants (CNVs) and presence/absence variants (PAVs) are crucial determinants of useful agronomical traits (Fig. 10.1). Both natural and artificial selection operate on these genomic variations to increase the overall fitness of the genotype and crop

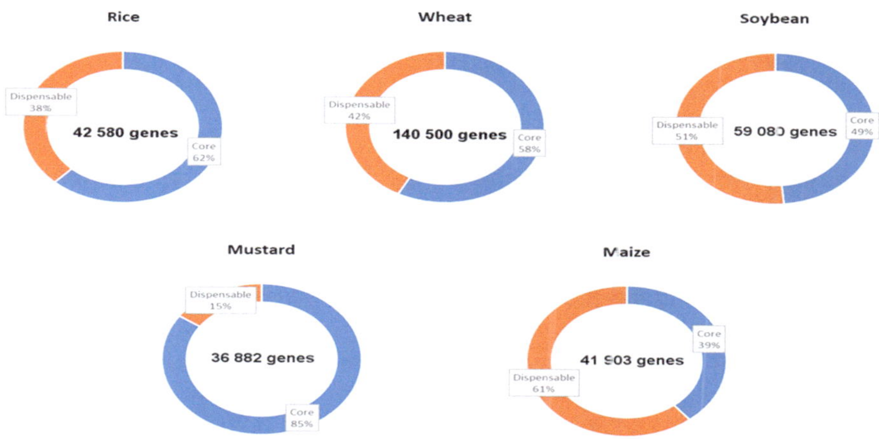

Fig. 10.2 The pan-genome of crop plants exhibits variable portions of core genes and dispensable genes. The pan-genome size of each species is indicated by the number of pan-genes in a particular species at the centre of the ring

productivity in an agricultural field, respectively. We depend on a single reference genome for understanding the existing genomic variations in a single species and this approach has a limited scope. The pan-genome concept was developed to capture large variations in gene content among individuals of a particular species. Thus, the entire gene repertoire of a species can only be described by the sequences obtained from multiple individuals belonging to a species. Recent pan-genomic studies on crop species like rice, maize, wheat, soybean, etc. have brought a paradigm shift in our understanding of crop biodiversity and their genetic improvement.

Pan-genome refers to the full set of genes in a taxonomic clade like a species. The pan-genome can be divided into two groups: core genes shared by all individuals in a species and dispensable genes only present in few individuals or present as singleton gene in a specific individual. The proportion of dispensable gene varies widely between different species of crops (Fig. 10.2). For example, rice plant has 8% of dispensable genes, whereas maize plant has 61% of dispensable genes. The percentage of dispensable genes in a species reflects the diversity among individuals of a particular species. A species with higher ploidy and outcrossing rate is likely to have larger pan-genome with a higher percentage of dispensable genes. Rice has a large pan-genome (48,098 genes) with 41% of dispensable genes. However, the number of new genes discovered in a pan-genome is likely to decrease with inclusion of new individual samples of a species. The sample size and selection of samples are key factors in a pan-genome study. Ideally, sample size should be large enough to realize the full expansion of the pan-genome. For example, 3010 rice accessions were sequenced to obtain the large pan-genome of rice. A genome browser, Rice Pan-genome Browser (RPAN) can be used as a search tool for the rice pan-genome derived from this study and for visualization of all genomic variations as well. However, the expansion of the pan-genome in a species reaches a plateau after incorporation of finite number of genomes. The selection of sample has a significant impact on the size of the pan-genome. Thus, a pan-genome study should include both cultivable crop and its wild relatives to generate a large pan-genome with high content of dispensable genes. Crop species with a high ploidy level and a high outcrossing rate are likely to have a larger pan-genome with higher percentage of dispensable genes.

The core genes in a crop species are highly conserved and likely to play a critical role in essential cellular functions. On the other hand, dispensable gene contributes to the phenotypic diversity of agronomic traits in a crop species and has a critical role in enhancing its agricultural productivity. The rate of dispensable gene evolution is faster than that of core genes. Dispensable genes are enriched with adaptive functions such as gene regulation, signal transduction and responses to environmental and defence stimuli. Interestingly, dispensable genes have a high frequency of both synonymous and nonsynonymous substitutions and a high proportion of them is not functionally annotated in many crop species like soybean.

Structural variation is a major causative factor in generating dispensable genes. Structural variations are defined as the large genomic alterations more than 50 bp in length including duplications, deletions, insertions, inversions and translocations. Transposable elements (TEs) or jumping genes are one of the most dynamic parts of the genome and generate structural variations in the form of PAVs in many crop species. Similarly, structural variations are also generated through recombination of non-allelic non-homologous sequences. The structural variations have an adverse impact on the fitness of a crop species. Moreover, both polyploidy and outcrossing rate provide a capacity in crops to tolerate the structural variations in the genome. A polyploid species has a multiple copies of essential genes that provide a buffer against the deleterious effects of structural variations. For example, a tetraploid species of mustard plants has a higher percentage (38%) of dispensable genes in comparison to a diploid species (19%). An outcrossing species is likely to have abundant structural variations conferring heterozygosity to tolerate the adverse effects of deleterious mutations. For example, maize crops which were outbred extensively have 61% of dispensable genes in their genome.

The genetic basis of agronomic traits is controlled by structural variations in crop plants. The underlying genomic variation in a crop species may be linked to an agronomic trait using quantitative trait loci (QTL) or genome-wide association studies (GWASs). The dispensable genome of a crop is enriched with genes associated with biotic and abiotic stress resistance. The structural variations in the genes involved in the abiotic and the biotic stress in crop plant are given in Tables 10.2 and 10.3. In addition, structural variations in some genes in crop plants are known to regulate the photoperiod sensitivity and flowering time. For example, a copy number variation at the *HvFT1* locus is associated with the variation of flowering time in barley. Even the whole plant architecture such as height of a crop plant can be reduced drastically by structural variations. In wheat plant, duplication of a gene (*Rht-D1b*) leads to reduction of plant height by more than 70%. Structural variations also influence the overall yield of the crop and other

Table 10.2 Structural variations in the genes involved in the abiotic stress

Type of SV	Gene	Crop	Trait
PAV	*Sub1A*	Rice	Tolerance to submergence
PAV	*Pup1*	Rice	Tolerance to phosphorus starvation
CNV	*MATE1*	Maize	Tolerance to aluminium
CNV	*FR-2*	Wheat	Tolerance to cold climate
CNV	*FR-H2*	Barley	Tolerance to frost
PAV	*GmCHX1*	Soybean	Tolerance to salinity
CNV	*Ppd-B1*	Wheat	Sensitivity to photoperiod

Table 10.3 Structural variations in the genes involved in the biotic stress

Type of SV	Gene	Crop	Trait
PAV	*Pikm1-TS*	Rice	Resistance to blast disease
PAV	*Pi21*	Rice	Resistance to blast disease
PAV	*qHSR1*	Maize	Resistance to head smut
PAV	*Lr10*	Wheat	Resistance to leaf rust
CNV	*Rhg1*	Soybean	Resistance to cyst nematode
PAV	*R1, ELR*	Potato	Resistance to late blight
PAV	*Yr36*	Wheat	Resistance to stripe rust

associated traits such as grain quality or fruit quality. For example, a 1212-bp deletion in *GW5* gene can alter the grain weight and width in rice plants. Elongated fruit shape in tomato is due to copy number variation of *SUN* gene. Since structural variations create differences in the gene content between different individual lines, the heterosis effect in hybrids may be due to passage of complementary genes from the individual parental lines. Structural variations might have played a significant role in the domestication of crop plants. For example, during domestication of maize plant, both increase in apical dominance and decrease in tiller number occurred due to insertion of transposable elements at the *tb1* locus.

Pan-genomic study of crop wild relatives will reveal the full landscape of biodiversity in the species and this untapped resource can be utilized for boosting the crop productivity. In fact, wild relatives of crop plant are being used for backcrossing during conventional breeding. This pan-genomic studies in the crop species can be linked to the QTL/GWAS to identify useful elite genes for further breeding strategies. Thus, a comprehensive genome resource of crop plants will be created through pan-genome studies in near future which can be further exploited by plant breeders for enhanced agricultural productivity.

Computational analysis of the pan-genome is prerequisite for fully understanding the genomic landscape of a crop species. The pan-genome of a species is computationally represented by a data structure known as coordinate system where all genetic variants are explicitly represented in form of a sequence coordinate graph. Although there are many available software for pan-genome analysis in prokaryotes, recent development of a software like Pangloss for pan-genome analysis in eukaryotes will be highly useful for a crop species. Pangloss is written in the Python language to characterize the pan-genome of eukaryotes for gene prediction, gene annotation and functional analyses. The total diversity in a crop species in form of a pan-genome can be visualized in form of a phylogenetic tree. A phylogenetic tree from whole-genome data can be reconstructed using both DNA sequences and the gene content. In a sequence-based phylogenetic tree reconstruction, genomic sequences from all variants are first aligned using a multiple sequence alignment and a phylogenetic tree is then reconstructed using the evolutionary distances. In gene content approach, the presence and absence of a gene in the genome is scored in binary numbers 1 and 0, respectively, followed by construction of a distance matrix consisting of 1 and 0 to represent total pan-genomic profile of a crop species. The distances between different variants are measured in terms of Jaccard distance and Manhattan distance. Manhattan distance refers to sum of absolute gene-wise differences between two genomes, whereas Jaccard distance between two genomes measure the amount of intersection in terms of presence or absence of a gene. If two genomic sequences are identical, both Manhattan distance and Jaccard distance are 0.0. Conversely, if there is no overlap between two genomic sequences, both measurements are 1.0. A distance-based phylogenetic approach such as neighbour joining or UPGMA will generate an evolutionary tree exhibiting relationship among different variants of the pan-genome in a crop species based on presence or absence of individual genes.

10.3 Assembly of Crop Genomes

Plants have evolved a large and complex genome for their survival in the terrestrial environment. The genome of the first plant species, *A. thaliana* followed by genome of some economically important species like rice, maize and papaya were sequenced using the Sanger method. The advent of next-generation sequencing technology has speeded up the pace of genome sequencing with a reduced cost. The second-generation deep sequencing of crops is challenging due to large genome size, duplications and repeat contents. The genome assembly process consists of a combination of sequencing and computation. The reads generated by sequencing is combined using a computer program called assembler. Therefore, assembly and analysis of short reads in plant genome needs substantial bioinformatics skill. The short reads generated through second-generation technology needs high coverage from 50X to 100X; even this high coverage of 100X may be inefficient in deciphering complex plant genomes. On the other hand, long-read technologies do not need such high coverage and a coverage of 20X to 30X may be sufficient to get a good assembly of the genome. Thus, second-generation deep sequencing technology has produced a draft genome of many plant species lacking almost 20% of the vital genomic information about the species. The resulted draft genome cannot be used reliably to understand the repetitive part of the genome and to discriminate between functional genes and pseudogenes as well.

In fact, a plant genome appears like a gene island surrounded by more than 80% of the repeat sequences. Transposons have played a profound evolutionary role in the evolution of plant species and transposable elements (TE) constitute the major portion of repetitive sequence in a plant genome. The small genome crops like rice have only 17% of transposons, whereas a large genome of maize have almost 85% of transposable elements. The abundant presence of repetitive transposons in a crop species poses a critical challenge in assembly of their genome. The short reads have less power to resolve the repeats in the plant genome. Thus, longer reads generated by single-molecule sequencing must be combined with short reads to resolve the repetitive DNA. The longer repeats than the read length create gaps during the assembly process which in turn can be resolved by paired-ends during sequencing.

Polyploidy or fusion of two or more genomes in a species has played a significant role in the evolution of wide diversity in plants. It is a result of either whole-genome duplication known as autopolyploidy or crossing of two species followed by duplication known as allopolyploidy. Genome duplications produce new genes with new functions and thereby generate novel phenotypes. The polyploid plants have better adaptive capability in an ever-changing environment due to their genomic plasticity. A majority of crop plants like wheat, potato, sugarcane and banana are polyploids. The redundancy due to presence of two or more copies of genes can adversely affect the accuracy of the genome assembly. Gene duplication is another major force for creating new genes in a genome in form of paralogs. It is difficult to distinguish alleles from paralogs in the genome assembly of natural heterozygotes. We always look for lineage-specific genes in a crop species for functional studies. Sometimes, these lineage-specific genes are simply artefacts of misassembles. These artefacts can be avoided by developing some novel algorithms to identify real lineage-specific genes. The development of de novo approaches in assembly such as de Bruijn graphs combined with Eulerian paths has facilitated assembly of plant genomes with long repeats. With the advent of third-generation single-molecule sequencing such as PacBio, the assembly of plant genomes with long reads having an average length of 10,000 bp can circumvent the problem of long repeats. However, the high error rate (5–15%) of this newer technology is a major constraint in the application to crop genomes.

10.4 Identification of Homeologs

The homeologous genes in plants are the products of allopolyploidy and have a common ancestry like other homologous genes. Homoeologs are the gene pairs derived by speciation and again recombined in a genome of a species by allopolyploidy. In simple words, we can define homeologs as the orthologs between the subgenomes of a species. The subgenomes are individual genomes in an allopolyploid species each inherited from a different ancestral species. Thus, homeologs have a combined evolutionary and functional features of orthologs (derived by speciation) and ohnologs plus paralogs (derived by duplication) (Fig. 10.3). The homeologous genes are expected to follow same order (collinear) between the ancestral genome and the descendant genomes. This is known as positional homeology like positional orthology and the homeologous genes maintain one-to-one relationship. The best bidirectional hits (BBH) approach works well in inferring

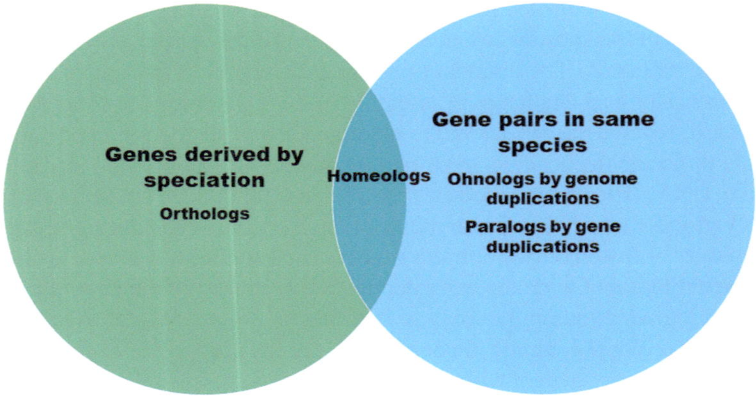

Fig. 10.3 Venn diagram showing relationship of homeologs with orthologs, ohnologs and paralogs. Homeologs are located at the intersection of orthologs and ohnologs plus paralogs

one-to-one relationship between orthologs. However, this approach fails to identify those homeologs underwent genomic rearrangements through single-gene duplication or translocation. The sequencing of complex crop genome is challenging due to difficulty in correct identification of homeologous sequences.

Computational analysis of the genome of an allopolyploid crop species such as wheat, cotton and coffee is necessary for accurate identification of homeologous genes. Evolution-based computational algorithms are used to infer homoeologs from the genome sequences. These algorithms are based on two approaches: the phylogenetic tree-based and the graph-based. The phylogenetic tree-based approach reconciliates gene tree with species tree to distinguish orthologs (speciation event) and paralogs (duplication event). It distinguishes each node of a phylogenetic tree either as a speciation or duplication event using species tree as a reference. Consequently, the gene pairs coalescing or joining at the speciation node and the duplication node are inferred as the orthologs and the paralogs, respectively. Ensembl Genomes has applied this algorithm to identify homoeologs in a polyploid species like wheat. Here, each subgenome is treated as a different species and orthologs between subgenomes are detected using phylogenetic tree-based orthology pipeline. Finally, the identified orthologs among different subgenomes are relabelled as homeologs in an allopolyploid species.

Graph-based orthology detection method is based on the similarity between homologous sequences. The best reciprocal hits are identified among two subgenomes using the BLAST to identify homoeologs. This approach was used successfully to identify homeologs in wheat. However, this method only detects one-to-one homeology but often fails to identify one-to-many and many-to-many homeologs. The limitation of this approach is more obvious for highly duplicated crop genomes. It generates many false-negatives because it does not account for the differential gene loss among the subgenomes. In addition, there are many fragmentary genes and sequencing errors in the crop genome leading to suboptimal score during the BLAST. Orthologous matrix (OMA) database has implemented another graph-centric approach for identification of homeologs in wheat. First, mutually closest homologs are chosen based on the evolutionary distance taking into account both differential genes loss and many-to-many relationship among genes. In parallel to earlier phylogenetic-based approach, orthologs between different subgenomes are detected using the standard pipeline followed by relabelling of orthologs as homeologs. Thus, the OMA approach is a better graph-based approach because it depends on the evolutionary distances in comparison to alignment score. However, the quality of assembly and annotation of allopolyploid crop genomes need to be improved for the high accuracy of homeolog inference.

10.5 Genomic Selection

Crop breeding depends upon the repetitive cycles of phenotypic selection followed by crossing in each generation to produce a superior genotype. The marker-assisted selection (MAS) was used for improvement of common crop in the recent past by detecting underlying major genes but failed to detect minor gene effects in the breeding population. With the availability of whole genomic sequences of various crop species, the minor-effect genetic variants associated with agronomic traits can be identified across the genome. The genomic selection operates on these genome-wide genetic variants circumventing the need for repeated cycles of phenotype selection. It selects the best candidates as parents for the next breeding cycle using predicted breeding value which includes their genotypes, their phenotypes and the genotypes of their relatives (Fig. 10.4). The breeding value of an individual is measured by the average performance of its progenies. Single-nucleotide polymorphisms (SNPs) are the variations at the nucleotide level in a population and are extensively used for identification of more than 10,000 quantitative trait loci (QTL) having economic importance. In a typical genomic selection program, there

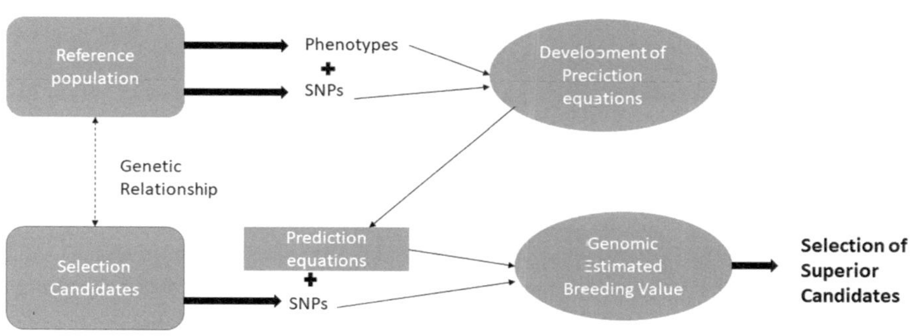

Fig. 10.4 Schematic diagram showing development of prediction equations from phenotype and genotype in form of thousands of SNPs in the reference population and subsequent application of prediction equations in the selection of candidates using thousands of SNPs data for computing genomic estimated breeding value

are two kinds of distinct but related population: the training population and the breeding population. The genotype with all genome-wide markers and phenotype of individuals in the training population are known, whereas only genotypes including all genome-wide markers of individuals in the breeding population is known without any knowledge of their phenotype. The genetic values of individuals in the breeding population need to be predicted during genomic selection. A prediction model is developed from the training population to predict the genomic estimated breeding values (GEBV) of individuals in the breeding population. This approach has a greater power in capturing the effects of small-effect loci as well. Overall, this model captures all the additive genetic variance of a particular trait of economic importance. Genomic selection has a high accuracy of genomic-enabled prediction for simple traits with high heritability. Moreover, it also has a potential to improve the complex traits with low heritability. Thus, genomic selection can play a significant role in enhancing genetic gain per unit time and cost in a breeding population through accelerated breeding cycles.

Since the variance of a complex trait is assumed in various forms in each type of prediction model, the field response of different types of prediction models varies widely. The general model for the whole-genome regression analysis can be formulated as:

$$y = Xb + Wa_m + e$$

where y is the vector representing phenotypic value,

X is an incidence matrix for the fixed effects,

b is a vector of fixed effects including the overall mean (μ),

W is a matrix of genotypic score of each SNP marker,

a_m is the vector representing the additive effects of the SNP marker

and e is the vector of random residual effects with distribution $e \sim N\ (0,\ I\sigma_e^2)$.

The SNP genotypes at each locus are represented as a matrix of minor allele frequency (0, 1 and 2 copies) or deviation of minor allele counts from heterozygous genotype (1, 0 and −1) for diploid genotypes of AA, AB and BB, respectively.

The most common predictive models for quantitative traits are the genomic best linear unbiased prediction (G-BLUP) and the ridge-regression BLUP (RR-BLUP). The best linear unbiased predictor (BLUP) is the best model as it finds the maximum correlation between the predicted value and the true value. Further, it is linear because of linear combinations of estimation of model effects. It is also called as unbiased because both expected value and true value are equal. A crop breeder typically uses BLUPS for marginal effect predictions. In simple words, the BLUP is a linear combination of effect estimates from mixed model with random effects. The G-BLUP model is a mixed linear model estimated from genome-wide markers and a modified form of conventional BLUP model emphasizing on the pedigree relationship. This model estimates random effects and utilizes the genomic information from relatives giving more weights to closest relatives. The G-BLUP model can be formulated as:

$$y = Xb + Zu + e$$

where y is the vector of phenotypic observations,

b is the n x 1 vector of fixed effects,

X is the n x p design matrix relating observations to fixed effects,

Z is an n x n identity matrix connecting the n observations to the n individual effects,

u is the n x 1 vector of breeding values of each individual,

and e is the n x 1 vector of residual errors having the variance $I\sigma_e^2$.

More complex models can be developed in case of a plant breeding experiment. For example, y may represent phenotypes of crops and the g families or lines that are replicated over s environments. The Z matrix will be transformed to n X $(g + s)$ in such a case. The u represents the G-BLUPS or genomic estimated breeding value (GEBV) of all individuals in the matrix. G-BLUP model can be fitted with the package regress and synbreed in the R environment.

Since the number of SNP predictors is larger than the number of phenotypic observations during genomic selection, the regression coefficients are constrained by adding a small constant lambda. This process of shrinkage or regularization is known as ridge regression. The ridge regression (RR) is like least square method which shrinks the estimated coefficients towards zero. Here, lambda is a tuning parameter for penalty which varies between 0 and infinity. The bias in the model increases with the amount of shrinkage (large lambda) but there is a corresponding decrease in the variance. RR-BLUP is also a mixed linear model where variance is treated as equal for all the markers with small effects. RR-BLUP model can be developed in the R environment using a package called solving mixed model equations in R (sommer) and rrBLUP.

Bayesian analysis is often conducted using Markov Chain Monte Carlo (MCMC) methods. Gibbs sampler is one of the most common Bayesian method for genomic selection. Bayesian whole-genome regression prediction models can be developed using the BGLR package in the R environment. It implements a variety of Bayesian approaches including Bayesian ridge, Bayesian Lasso, BayesA, BayesB, BayesC, etc.

10.6 Crop Phenomics

The plant phenotyping is a core concept of plant breeding where some elite plants are selected for subsequent crossing and genetic gain. This traditional phenotyping method focusses on one or few traits in a particular environment and often leads to a marginal annual genetic gain (0.5–1%) in the productivity of major cereal crops. The crop phenomics deals with collection of multi-dimensional phenotypic data at various scales such as cell level, organ level, plant level and population level (Table 10.4). There are automated phenotyping platforms such as for generating high-throughput data from individual plants. The most common phenotyping indices of individual plants are leaf length, leaf area and fruit volume.

Table 10.4 Phenomics traits in crops

Morphological traits	Structural traits	Physiological traits	Performance traits
Plant volume	Cell division rate	Cell turgor	Biomass/hectare
Stalk shape	No. of vascular bundle	Water transportation	Seed yield
Coverage fraction	Vessel size	Stalk lodging	
Internode diameter	Cross section thickness	Interception of PAR	

A genomic selection model can be applied for genetic gain under variable season or field condition. However, combination of genomics data with phenomics data obtained from repeated experiment in variable environment has a better potential for genetic gain in crop breeding. In phenome to genome approach, the SNPs and genomic regions are associated with agronomical phenotypic traits using high-throughput phenotyping tools. The final statistical model consists of three components: genotype, environment and management (G X E X M).

Image-based phenotyping is used frequently in both laboratory and field environments. Plant phenotypes are divided into four groups: colour, texture, quantity and geometry in image-based phenotyping. Each plant is exposed to top and side cameras at three different wavelength bands: visible, near-infrared and fluorescence to get thousands of images during whole phenotyping period. The visible range provides colour images regarding nutrition, health, growth and biomass status of an individual plant. Moreover, infrared range detects a quantitative measure of the water content and fluorescence imaging senses chlorophyll and other fluorophores in a plant. Machine learning and deep learning techniques are widely used in the image-based phenotyping such as object detection and classification. The phenotype image analysis consists of four successive steps: pre-processing, segmentation, feature extraction and post-processing. The first step involves preparation of the image for analysis like detection of outlier, reproducibility of trait and normalization of phenotypic profile. It is followed by the second step of segmentation dividing the image in foreground (the plant) and background (the imaging chamber) parts. Feature extraction selects the optimal set of explanatory variables in a stepwise manner using variance inflation factors to get a list phenotypic traits. In general, about 400 phenotypic traits are extracted from the image of each crop plant. The final step of post-processing is summarization of all computed results.

10.7 Crop Systems Modelling

Agricultural systems science is an interdisciplinary area to understand the overall behaviour of complex agricultural systems in different time and space. An agricultural system model consists of components and their interactions with respect to natural resources, agricultural production and human factors. These models can predict the overall performance of agro-ecosystem in near future. They are also helpful to land managers and policymakers in taking decisions regarding financial planning, crop management, land management and pest management. These software programs are known as decision support systems. First crop systems models were statistical models to predict the response of a system such as crop yield based on the past data sets. However, statistical models are not suitable to predict the impact of unseen climate changes such as increase in CO_2 concentration in the atmosphere. Statistical models are generally useful in case sufficient historical data sets are available for prediction.

The dynamic system simulation models are widely used to understand crop and farming system models in response to certain external changes such as weather or management practices. The typical output of a dynamic model is daily outputs of a specific crop over a period of time. The models are highly complex having numerous descriptive variables and parameters with long run times. Thus, summary models are derived from a complex model suitable in certain situations. A common model of crop system, Agricultural Production System Simulator (APSM) is a complex dynamic model, which predicts the yield of a crop as a function of time and space based on several inputs such as weather and soil properties. This model can be implemented in the R environment using a package called apsmr. Similarly, World Food Studies (WOFOST) crop growth simulation model and Light interception and utilization (LINTUL) can also be implemented in the R using the packages Rwofost and LINTUL-package, respectively. AquaCrop is a mechanistic crop growth generic model from emergence to maturity developed by Food and Agriculture organization. It is most widely used generic model to simulate growth of different crops under variable climates and geographical locations. AquaCropR is an R package built upon the AquaCrop software with some additional functions. An R package for agricultural data sets known as agridat is available in the R environment.

Box 10.1 Heritability

The heritability in broad sense is a measure of relationship between phenotypic value and the breeding value for a particular trait in a plant or animal population. Breeding values are likely to have more impact on a phenotype if heritability of the trait is high. On the other hand, heritability in the narrow sense is the square of correlation between breeding value and phenotypic value. The value of heritability varies from zero to 100% and traits having heritability more than 0.4 are treated as highly heritable. For example, the traits for percent fat and percent protein in milk of dairy cattle have a heritability of 0.5 and are highly heritable. It is a measure of overall population and varies between different populations and environments.

Box 10.2 Breeding Value

The plants or animals with best genes are believed to have the best breeding value. In artificial selection, the parents with best breeding value are chosen to

(continued)

Box 10.2 (continued)
contribute genes to the next generation. The relationship between a phenotype and breeding value is measured in terms of heritability. The phenotype is the correct manifestation of underlying breeding value in case the heritability of the trait is high. Conversely, the low heritability indicates little connection between a phenotype and its breeding value. The breeding value is predicted from objective and numerical phenotypic and genomic data. Sometimes, we rely on the information from relatives to achieve high accuracy in prediction. Accuracy measures the parity between true breeding value and its predicted value.

10.8 Exercises

1. The R package ZeBook contains a dynamic model of crop growth for maize cultivated in potential conditions. Three state variables, namely leaf area index (LAI), total biomass (B) and cumulative thermal time since plant emergence (TT) indicate the overall crop growth. Find the parameters and compute the growth of crop in terms of three state variables from day 100 to 150. Plot the increase in total biomass during this period.

Solution (Figs. 10.5 and 10.6)

```
library(ZeBook)
> weather = maize.weather(working.year=2010, working.site=30,weather_
all= weather_ EuropeEU)
> maize.define.param()
        Tbase  RUE    K   alpha   LAImax TTM    TTL
nominal   7    1.85  0.7  0.00243   7     1200  700
binf     6    1.50    0.6   0.00200   6   1100    600
bsup     8    2.50   0.8    0.00300   8  1400    850

> model<-maize.model2(maize.define.param()["nominal",], weather, sdate=
100, ldate=150)
>model

>plot(model$B,type="o",pch=16)
```

2. The alpha lattice design of spring oats is an available dataset with the R package *agridat*. This dataset contains 72 observations on five variables, namely plot number (plot), replicate (rep), incomplete block (block), genotype (gen) and dry matter yield (yield). Estimate the genetic effects using the best linear unbiased prediction (BLUPs) and heritability for yield in the R environment.

	day	TT	LAI	B
1	100	0	0.01	1
2	101	5.7	0.01096838	1.280034
3	102	8.85	0.01155406	1.570097
4	103	10.9	0.01195722	1.89797
5	104	13.9	0.01256635	2.158567
6	105	18.6	0.01356919	2.448584
7	106	22.5	0.01446761	2.83334
8	107	27.35	0.0156587	3.263947
9	108	33	0.01736024	3.741912
10	109	40.3	0.01928585	4.272062
11	110	47.95	0.02178854	4.649134
12	111	55.45	0.02455956	5.307182
13	112	62.35	0.02743198	6.000904
14	113	69.6	0.0308017	6.437204
15	114	71.6	0.03184406	7.218537
16	115	79.4	0.03605087	8.193251
17	116	87.1	0.04074839	9.129116
18	117	93.95	0.04546889	10.315255
19	118	102.15	0.05176956	11.70617
20	119	112.7	0.06099118	13.325867
21	120	122.7	0.07127538	14.849096
22	121	130.05	0.08009575	16.559796
23	122	136.85	0.08925426	18.244323
24	123	142.3	0.09742308	20.653345
25	124	146	0.1034692	21.983894
26	125	149	0.10867119	24.622408
27	126	152.2	0.11449455	27.91529
28	127	156.9	0.12349833	30.83622
29	128	161.95	0.13391973	33.379546
30	129	167.75	0.14687922	36.210873
31	130	172.9	0.15947608	37.538354
32	131	180.5	0.17962282	41.487959
33	132	185.9	0.19560852	44.089083
34	133	189.2	0.20637655	48.044265
35	134	192.5	0.21761955	52.448912
36	135	197.1	0.23411804	58.042912
37	136	201.7	0.25182415	64.0832
38	137	207.6	0.27618779	69.644344
39	138	214.65	0.30800158	77.872254
40	139	221.85	0.34406339	86.626852
41	140	230.4	0.39164291	97.317953
42	141	239.35	0.44793056	109.694301
43	142	249	0.51675184	123.686229
44	143	260.6	0.61118813	139.521234
45	144	273.4	0.73264198	157.873602
46	145	287.55	0.8905263	179.101874
47	146	299.7	1.05115868	192.480004
48	147	309.2	1.19551326	209.738054
49	148	318.6	1.35402968	228.302225
50	149	328.75	1.54257628	256.513084
51	150	337.1	1.7133917	268.973687

Fig. 10.5 Growth of maize crop in terms of three state variables

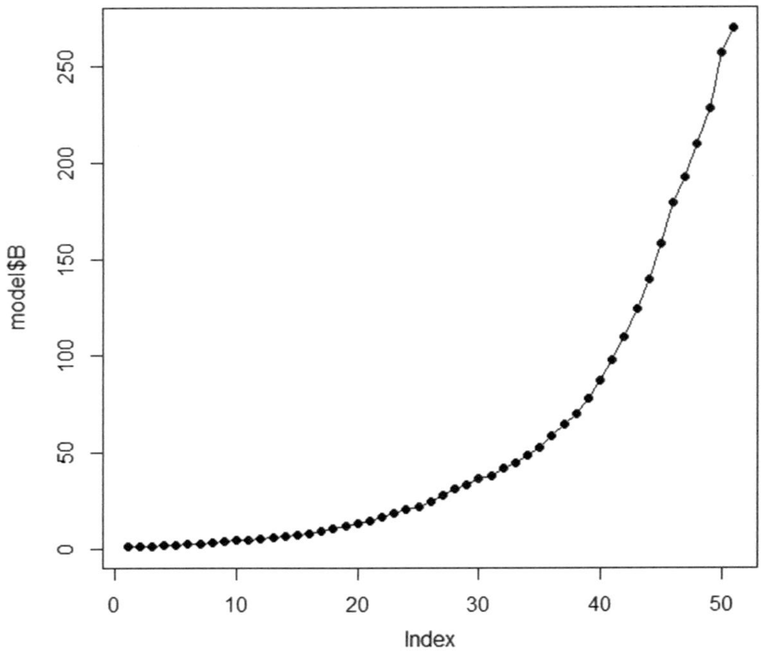

Fig. 10.6 Increase in total biomass from day 100 to day 150

```
> library(agridat)
> library(lme4)
> library(emmeans)
> data(john.alpha)
> dat <- john.alpha
> model <- lm(yield ~ 0 + gen + rep, data=dat) # Randomized Complete Block
(RCB) design
> model1 <- lm(yield ~ 0 + gen + rep + rep:block, dat) # Intra-block analysis
> model2<-lmer(yield~0 + gen+rep+ (1|rep:block),dat)# Combined inter-intra
block analysis
> anova(model2)
Analysis of Variance Table
    npar  Sum Sq  Mean Sq    F value
gen  24   380.44  15.8515    185.9959
rep  2    1.57    0.7851     9.2124
> means <- data.frame(rcb=coef(model)[1:24],
+ ib=coef(model1)[1:24],
+ intra=fixef(model2)[1:24]) # Variety means
> head(means)
       rcb      ib         intra
genG01 5.201233 5.268742 5.146433
```

```
genG02 4.552933 4.665389 4.517265
genG03 3.381800 3.803790 3.537933
genG04 4.439400 4.728175 4.528828
genG05 5.103100 5.225708 5.075944
genG06 4.749067 4.618234 4.575395
> covtosed <- function(x){
+ n <- nrow(x)
+ vars <- diag(x)
+ sed <- sqrt( matrix(vars, n, n, byrow=TRUE) +
+ matrix(vars, n, n, byrow=FALSE) - 2*x )
+ diag(sed) <- 0
+ return(sed)
+ } # conversion of variance-covariance matrix to SED matrix
> model5blue <- lmer(yield ~ 0 + gen + rep + (1|rep:block), dat) # BLUE
> ls5blue <- emmeans(model5blue, "gen")
> con <- ls5blue@linfct[,1:24] # contrast matrix for genotypes
> tmp <- tcrossprod( crossprod(t(con), vcov(model5blue)[1:24,1:24]), con)
> sed5blue <- covtosed(tmp)
> tmp <- tcrossprod( crossprod(t(con), vcov(m5blue)[1:24,1:24]), con)
> sed5blue <- covtosed(tmp)
> vblue <- mean(sed5blue[upper.tri(sed5blue)]^2) #average variance of
difference between genotypes
>model5blup<-lmer(yield~0+(1|gen)+rep+(1|rep:block), dat) #best linear
unbiased prediction (BLUP) for various effects in the model
> re5 <- lme4::ranef(model5blup,condVar=TRUE)
> vv1 <- attr(re5$gen,"postVar")
> vblup <- 2*mean(vv1)
> vblup
[1] 0.0577334
> sg2 <- c(lme4::VarCorr(model5blup)[["gen"]]) #h^2 (heritability)
> h2<-sg2 / (sg2 + vblue/2)
>h2
[1] 0.8030173
```

10.9 Multiple Choice Questions

1. The structural variants associated with agronomic traits are:
   (a) CNVs
   (b) PAVs
   (c) CNVs and PAVs
   (d) None of the above

2. The percentage of dispensable genes in maize is:
   (a) 51%
   (b) 61%
   (c) 71%
   (d) 81%
3. The phenotypic diversity of crop plants are due to the presence of:
   (a) core genes
   (b) dispensable genes
   (c) pan-genes
   (d) All the above
4. The amount of intersection in terms of presence and absence of a gene between two genomes is measured by:
   (a) Jaccard distance
   (b) Manhattan distance
   (c) Robinson–Foulds distance
   (d) None of the above
5. The next-generation sequencing in crops is challenging due to:
   (a) Large genome size
   (b) Duplication
   (c) Repeat content
   (d) All the above
6. The percentage of transposable elements in maize crop is:
   (a) 75%
   (b) 85%
   (c) 90%
   (d) 95%
7. Homeologous genes in crops are the product of:
   (a) Autopolyploidy
   (b) Allopolyploidy
   (c) All the above
   (d) None of the above
8. The predicted breeding value of a parent crop consists of:
   (a) Genotype
   (b) Phenotype
   (c) Genotype of relatives
   (d) All the above
9. The genomic estimated breeding values of an individual capture:
   (a) Additive genetic variance
   (b) Dominance variance
   (c) Epistatic variance
   (d) None of the above

10. The common predictive model for quantitative traits is:
    (a) BLUP
    (b) BLPP
    (c) BLAP
    (d) BPLU

Answer: 1. c 2. b 3. b 4. a 5. d 6. b 7. b 8. d 9. a 10. a

Summary

- The structural variants in crops such as copy number variants (CNVs) and presence/absence variants (PAVs) are crucial determinants of useful agronomical traits.
- The percentage of dispensable genes in a crop species reflects the diversity among individuals of a particular species.
- Structural variation is a major causative factor in generating dispensable genes in crops.
- The dispensable genome of a crop is enriched with genes associated with biotic and abiotic stress resistance.
- The second-generation deep sequencing of crops is challenging due to large genome size, duplications and repeat contents.
- The homeologous genes in plants are the products of allopolyploidy.
- Genomic selection has a high accuracy of genomic-enabled prediction for simple traits with high heritability.
- The most common predictive models for quantitative traits are the genomic best linear unbiased prediction (G-BLUP) and the ridge-regression BLUP (RR-BLUP).
- The crop phenomics deals with collection of multi-dimensional phenotypic data of crops at various scales using automated platforms.
- The dynamic system simulation models are widely used to understand crop and farming system models in response to certain external changes.

Suggested Readings

Wallach D, Makowski D, Jones J, Brun F (2013) Working with dynamic crop models. Academic Press, Amsterdam

Isik F, Holland J, Maltecca C (2017) Genetic data analysis for plant and animal breeding. Springer, Cham

Normanly J (2012) High-throughput phenotyping in plants: methods and protocols. Humana Press, Totowa

Busch W (2017) Plant genomics: methods and protocols. Humana Press, Totowa

Farm Animal Informatics

11

Learning Objectives
You will be able to understand the following after reading this chapter:

- Role of bioinformatics in animal husbandry and veterinary science.
- Modelling of farm animal systems.
- Nutritional models of important farm animals.
- Computational model of laying hen.
- Lactation model of a dairy cow.
- Livestock genomics.
- Livestock phenomics.
- Methods of genomic selection for improvement of farm animals.

11.1 Introduction

Farm animal production system is a complex process consisting of three interacting components: biology of the animal, environment and management practices. Thus mathematical modelling of this system using automated real-time big data may provide us new clues for enhanced efficiency in animal productivity. With the advanced sensor technology, a large amount of data on individual animals can be generated in a short span of time. The recent advancements in big data analytics and precision agriculture, new algorithms and data structures are needed to develop next-generation farm animal production models taking advantage of sensor data and artificial intelligence. Mathematical modelling of this complex farm system has been done from individual farm animals including many ruminant and non-ruminant animals to entire farming system. There are two kinds of common modelling approaches in the animal production system. The first model is an

B. K. Tiwary, *Bioinformatics and Computational Biology*,
https://doi.org/10.1007/978-981-16-4241-8_11

empirical model which is derived from experimental data to infer relationships among different components at a single level. For example, an empirical model can describe the relationship of enteric methane production with other explanatory variable such as daily feed intake and level of milk production. Moreover, these empirical models are merely based on observation and experiment but do not explain the underlying biological processes. With the advent of big data and advanced analytics methods in the animal production system, empirical methods can be a powerful tool to understand the physiological and metabolic processes involved in an animal production system. On the other hand, the mechanistic model is a process-based model with several individual components and their specific interactions with explaining the underlying structure of the system. A dynamic mechanistic model of rumen physiology, for example, consists of 19 state variables such as microbial biomass, carbohydrates, and fatty acids and is driven by inputs of various nutrients. These models need meticulous construction and in turn provide accurate estimates of system functions. Even, the system functions can be further extrapolated beyond the data points used for the model construction. Mechanistic models have been proved as a better model than the empirical model in predicting methane emission in cattle. However, this modelling approach is usually improved by incorporating other variables such as genetics, behaviour, environment and management of individual animals.

11.2 Whole-Farm Systems Model

The whole-farm systems model is an interdisciplinary collaborative effort between experimental biologists and computational biologists. The model development can be developed in five steps which are as follows:

1. Description of the underlying biological system using existing models and literature.
2. Information flowcharts and development of pseudocode.
3. Translation of pseudocode to codes by the programmers.
4. Submodel functions are evaluated by existing experimental datasets.
5. Submodules are incorporated in larger modules and in turn modules are incorporated into systems model.
6. Evaluation of systems model and outputs for user application in the animal farm.

Thus, this modular format consists of several modules and submodules with clear documentation, readable and readily adaptable in a new setup. Individual modules can be adapted and improved further in a particular farm condition without disrupting other modules of a system. For example, a whole-farm dairy systems model consists of four integrated biophysical modules of Animal, Manure, Soil and Crop, and Feed storage. In addition, there are three system balance modules of water, energy and economics. The Animal module consists of three primary submodules: Animal Life cycle, Nutrition & Production and Management & Facilities (Fig. 11.1).

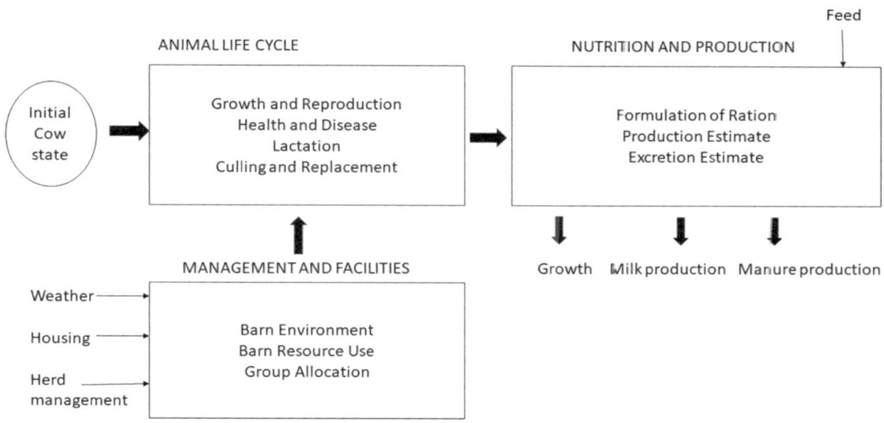

Fig. 11.1 Schematic diagram of an animal module with three submodules

The Animal Life Cycle submodule is a stochastic Monte Carlo model of events in life of a dairy cow from birth to culling or death. The Animal module can simulate stochastically the daily growth, production and reproduction of individual cows based on the inputs regarding feed, weather and management. The stochastic modelling simulates the probabilistic distribution of events for each individual cow in a herd accommodating interactions among cows. The Nutrition and Production submodule can predict the optimal diet of a cow for a desired milk production using linear programming. The Management and Facility submodule is a barn-level simulation including interaction of human intervention and environment with the cow herd. The Soil and Crop module simulates soil temperature, hydrology, nutrient dynamics, pasture growth and animal feeding using various climate inputs.

Interestingly, the feed storage module simulates the nitrogen and carbon loss during harvesting and storage of animal feeds. Finally, the Water Balance module is an all-encompassing module gathering information from four biophysical modules with water use and generation in cattle management.

11.3 Nutritional Models of Farm Animals

11.3.1 Dairy Cattle Model

The efficiency and robustness of farm animals has been a focus of attention in the field of animal nutrition. Productive animals are known to have better feed efficiency and as a result have margin over feed cost. Although selection can improve significantly the genotype of a farm animal in respect of some selected traits, other non-selected traits remain prone to be fragile under fluctuating environmental perturbations. The efficiency of a system is defined as the ratio between the fluxes of outputs and inputs. For example, feed efficiency of a cow refers to production of milk and corresponding supply of nutrients. Robustness is the ability of a living system to maintain its life trajectories under the influence of external and internal perturbations. The efficiency and robustness of a ruminant animal is regulated by two types of interacting regulations: homeorhetic regulations and homeostatic regulations. Homeorhetic regulations control the basic life processes associated with reproduction and growth of an animal such as gestation and lactation period of a cow. On the other hand, homeostatic regulations operate on the adaptive changes in animals under altered nutritional environment. The intake of dry matter and feeding behaviour of animal was studied using regression models to assess the individual variations in feed intake or residual feed. Most of the mechanistic models developed, so far, have been focussed on the rate of feed intake and bite mass. Feed cost constitutes about 60–70% of the overall production costs in cattle production.

The rumen provides a complex ecosystem for the rapid growth and proliferation of billions of microbes. Many mechanistic models of ruminant digestion have been developed to simulate the major alterations in digestive efficiency. An empirical model of digestion showed that endogenous protein faecal losses is predominant in dairy cows in comparison to basic maintenance requirements in terms of body protein turnover. Another model measures the non-productive energy losses in dairy cows which adversely affect the milk yield. The rumen protein balance is the difference between the crude protein fluxes ingested and delivered to the duodenum. This balance reflects an equilibrium state between the recycling of nitrogen both through rumen wall and saliva and absorption of excess ammonia in the rumen. The variations in rumen protein balance reflect the overall protein efficiency of the individual animal. However, the genetic control of these variations and the range of individual variation are yet to be known.

The efficiency of a farm animal is measured experimentally by the metabolic partitioning of the total energy between maintenance energy and other energy expenditures. For example, in a milk producing animal like dairy cow or goat, partition of carbon occurs in favour of milk and in disfavour of carbon related to carbon dioxide, faecal, urinary and methane. Thus mechanistic modelling of farm animals needs a multidisciplinary team consisting of computational biologists, nutritionists, geneticists, reproductive physiologists and ethologists.

11.3.2 Pig Model

The precision livestock farming is a management principle based on the processes and technology of process engineering. The precision feeding is a part of this system and requires proper amount of feed with appropriate composition in a timely manner to an individual animal to enhance overall farm productivity. Like cattle farming, pig farming has a major share of feed cost. But, the efficiency of conversion of nutrients to animal product is very low in farm animals. For example, the nitrogen (protein) retention efficiency varies from 15% to 33% in growing pigs. The nitrogen which is not retained by the animal is excreted in the external environment and causes nitrate pollution resulting in algal blooms in the water bodies. The tailoring of diet for individual animals as per production objective is a key principle in precision feeding. The precision feeding of pigs can be implemented by collecting data automatically from individual animals, data processing and actions required to control the system. The feed intake and body weight data of growing pigs is estimated on real-time basis for individual animals. A mathematical model can be developed to estimate the real-time pig individual nutrients requirement. For example, the daily concentration of lysine in the diet is estimated from the feed intake and body weight of an individual pig. This data is used to develop both empirical and mechanistic models. Empirical model predicts the expected feed intake, weight gain and body weight for the subsequent day, whereas mechanistic model computes optimal concentration of lysine in feed each day using all three variables estimated using empirical model. Based on the results of mathematical modelling, right amount and composition of feed is provided to each growing pig for optimum productivity.

11.3.3 Sheep and Goat Model

Although sheep and goat are small ruminants, different feeding strategies are needed in these small ruminants due to their specific physiological requirements. For example, the primary production of dairy sheep is wool in addition to milk production. Sheep and goat eat more food as a percentage of body weight for regular maintenance. Thus, a high producing dairy sheep and goat have a feed intake of 4% to 7% of body weight but it never exceeds 4% of body weight in case of dairy cow. Production efficiency of an animal is maximization of its product such as milk or meat in comparison to inputs used and usually represented by the feed efficiency. Feed conversion ratio is generally reciprocal of the feed efficiency. The higher value of feed efficiency is desirable for an animal. On the other hand, lower feed conversion efficiency is preferred during an animal production process. Various nutritional models have been developed to enhance the production efficiency of goats and sheep. The current nutritional models are more comprehensive having many mechanistic components such as animal, dietary and environmental variables. The energy and nutrient requirement of sheep and goats can be predicted more accurately under diverse environmental and management practices. These nutrient models can be further improved by real-time monitoring of small ruminants along with environmental and production variables using modern sensor technology.

11.3.4 Laying Hen Model

A laying hen has a potential to produce at least 300 eggs annually. The weight of an egg is determined not only by the age and genetic potential but on the nutrients fed during laying period. The amino acid and energy requirement of a laying hen is necessary to predict the potential body weight at first laying and subsequent potential egg output over a time period. Energy constitutes about 70% of the costs incurred on the poultry feed. Methionine is the primary limiting amino acid in the feed of a

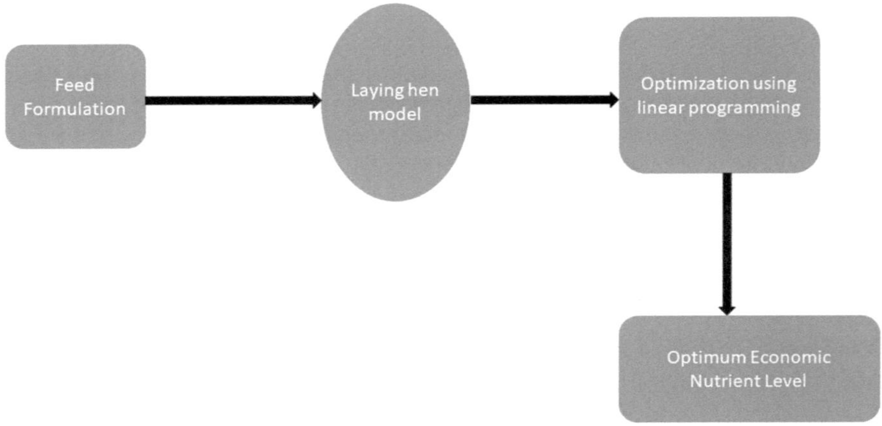

Fig. 11.2 Determination of optimum economic nutrient level for laying hen using optimization

laying hen. The age and body weight of an individual first laying bird determines its future laying performance in terms of egg number and egg weight. This characteristic feature of growth can be manipulated using different lighting and nutritional regimes. The change in photoperiod especially the initial and final photoperiod has a strong influence on the gonadal maturity of a laying hen. The daily intake of amino acids and energy is largely used by a laying hen for maintenance. The body protein content of a laying hen is found to be maximal at the age of sexual maturity and remains comparatively stable throughout the entire laying period. A mathematical model can predict food intake of a hen based on its body protein weight and potential daily egg output. The model computes the energy requirements of maintenance and egg output. However, the potential growth and egg output differ markedly between growing and reproducing birds. An individual hen responds linearly to the increasing amount of a limiting nutrient to its maximum genetic potential. The response to nutrients are represented in terms of an economically important output such as egg output. However, such responses are generally curvilinear when applied to a group of birds. A simulation model can provide the answer for optimum economic nutrient level for a group of laying birds considering marginal costs incurred and revenue generated (Fig. 11.2). First, a feed formulation is passed to a laying hen model to simulate the performance of hen. The costs and revenues are also calculated. During the optimization process, feed formulations are altered following certain rules. This optimization process iterates several times till it improves the value of objective function using linear programming. Thus, egg producers can maximize their profits by relying on this optimization process. In addition, it excludes the necessity for an expensive and tedious long-term laying trial to measure the response of laying hens to various feeds with differential nutrients.

11.3.5 Lactation Model

Lactation in a cow is a product of specialized epithelial cells of the mammary gland. These cells undergo variation in number and their secretory activity. The proliferation of cells starts at very beginning of pregnancy and reaches its maximum before parturition. The shape of a lactation curve is characterized by an ascending phase after parturition reaching its maximum peak followed by a declining slope till the dry-off of the dairy cattle (Fig. 11.3). A lactation curve in dairy cattle consists of two distinct phases: first phase of increasing milk production up to the peak level followed by second phase of decline of milk production. Primarily, lactation is a deterministic, continuous and regular process. However, stochastic components of lactation curve are genetic variability among animals and their health status, feeding and environmental practices, and overall farming system. The mathematical model of lactation curve is described by analytical function of time, where y_t is the daily milk production recorded at time t:

$$y_t = f(t)$$

There are many empirical models used for fitting the lactation curve data. Early models emphasized on the deterministic component of the lactation curve. The

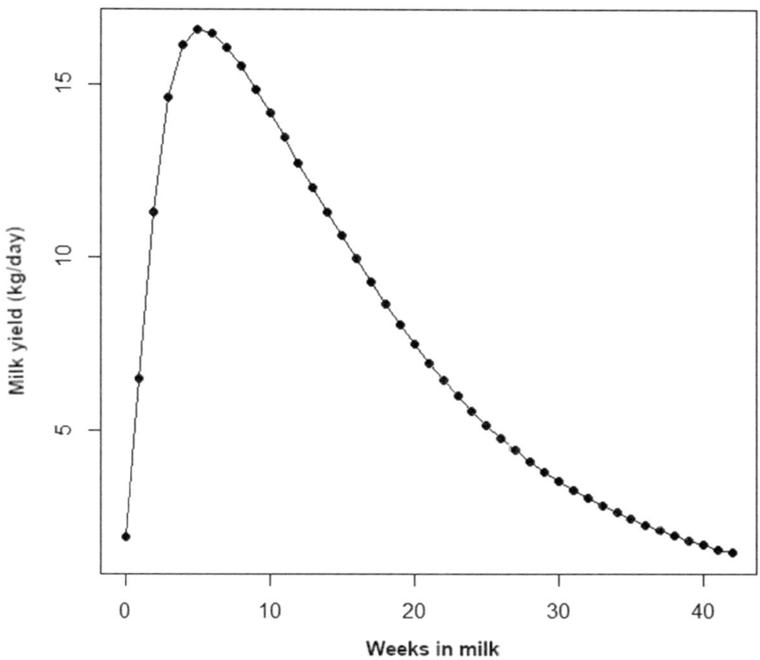

Fig. 11.3 The lactation curve of a dairy cattle showing ascending, peak and declining phases

Wood incomplete gamma function is a popular empirical model of the lactation curve. The function of this model is $y_t = at^b e^{ct}$ with three parameters a (a scale tends to increase with parity), b (controls the rate of increase to lactation peak) and c (rate of decline after the peak).

11.4 Livestock Genomics

A high-quality annotated reference genome of a farm animal is not only prerequisite for discovery and analysis of genetic variation but also useful in connecting genotype to function. The discovery of molecular genetic variants in the genome sequences and subsequent development of single-nucleotide polymorphism (SNP) chips are necessary to understand the genetic basis of complex traits such as growth, body composition, feed conversion, reproduction and response to a microbial infection. The genomes of farm animals have undergone a change in terms of removal of disease genes under the influence of purifying selection during successive selective breeding of superior animals. The genomics-enabled genetic improvement of farm animals is known as genomic selection. An overview of genomic resources available for farm animals is listed in Table 11.1. The reference sequence of majority of farm animals can be accessed on the Ensembl genomic database. Chicken is an important food animal constituting 41% of meat produced in the world. The draft genome of chicken was released in 2011 by the International Chicken Genome Consortium. It consists of 31 chromosomes and 14,093 unplaced scaffolds. The polymorphisms related to quantitative traits present in the chicken genome can be investigated and analysed for directed evolution of superior breeds. A draft reference genome of pig was developed in 2017 by the Swine Genome Sequencing consortium. It consists of 20 chromosomes including two sex chromosomes along with 583 unplaced scaffolds. Since a reference genome does not represent complete genetic diversity in the pig genome due to divergence in different geographic regions, genome sequences of 12 pig breeds were also sequenced. The bovine genome consists of 30 chromosomes built from 2211

Table 11.1 Genomic databases of farm animals

Database	Description	URL
Ensembl genome browser	Genomic database of all farm animals	www.ensembl.org
National Animal Genome Research Program	A comprehensive resource of USDA containing animal quantitative trait loci database (AnimalQTLdb) and animal trait correlation database (corrDB)	www.animalgenome.org
Bovine genome database	A database of bovine genome based on Hereford cow.	bovinegenome.elsiklab.missouri.edu
Chickspress	A database for chicken gene expression	http://geneatlas.arl.arizona.edu

scaffolds. Bovine Genome Database (BGD) is a web-based resource providing access to bovine genome assemblies and annotations. The domestic goat genome sequence was developed by the Bangladesh Goat Genome Consortium in 2019. It consists of 29 chromosomes assembled from 3972 scaffolds. The reference genome of sheep consists of 27 chromosomes assembled from 2641 scaffolds. AnimalQTLdb is a database having QTL, candidate gene, GWAS data and copy number variations (CNV) mapped to various farm animal genomes. Online Mendelian Inheritance in Animals (OMIA) has a manually curated collection of inherited disorders and genes in various species of farm animals.

11.5 Livestock Phenomics

The phenotype is the measurement of some features in an animal and has been a basis of all genetic improvement. High-throughput phenotyping is very crucial in closing the gap between genotype and phenotype. The real-time acquisition of high dimensional phenotypic data such as physiological or behavioural associated with production on individual animal scale is known as livestock phenomics. The livestock phenomics is more challenging than the plant phenomics because animals have a longer generation interval than the crops and they change their location frequently unlike crops in a field. Phenomics data can be measured at different levels from molecular level to morphological level. Molecular measurements include transcripts, proteins and metabolites expressed in different cells or tissues at different time points. Morphometric, behavioural and physiological data are other higher levels of phenomics data. Recent developments in sensor technology have given us opportunity to measure difficult or previously unmeasurable phenotyping traits in farm animals. The knowledge of most appropriate phenotype such as feed efficiency in a grazing livestock system will give deep insights into the biology involved in manifestation of this phenotype. For example, interactions between genotypes and environment and pleiotropic effects of genes can be well understood using phenomics approaches. A sensor mounted on an animal allows comprehensive two- and three-dimensional measurements or images of animal behaviour in a pasture. Physiological traits such as body temperature, heart rate, respiratory rate and rumen function can be monitored at a real-time basis on each individual animal. Sometimes, radiant temperature in the farm environment are monitored using thermal cameras. Many phenotypes of economic importance in animals such as feed conversion efficiency, disease resistance and reproductive potential are extremely difficult to measure in a large pastoral environment. These complex traits are derived from individual components and their interactions. The rapid generation of multisensor high-throughput phenotypic data in an animal farm over time poses a grand challenge in big data management. Further, the computational analysis of these big phenotypic data for biological interpretation is necessary to enhance the farm productivity. In dairy industry, the performance of a Holstein breed animal in the USA is measured by 42 traits of economic importance. The success of a genomic selection program solely depends upon large number of phenotypic measurements.

11.6 Genomic Selection in Farm Animals

Recent advances in genome sequencing technology has resulted in availability of reference genome sequences for most livestock species. The single base pair variations of individual animals from the reference genome called single-nucleotide polymorphisms (SNPs) are genotyped as genetic markers using SNP-chip based genotyping technology. SNPs are abundant markers usually located outside the genic regions across the entire genome. They are linked to the genes or genic regions known as quantitative trait loci (QTL) through linkage disequilibrium (LD). The LD measures the non-random association between the alleles across loci and is a ratio between expected and observed allele frequencies. In fact, it reflects the physical distance between two loci; close distance between loci indicates a strong LD justifying use of one locus as a proxy for the other one. The LD is strong for dairy cattle due to small effective population size (N_e) and often broken down by recombination events. A bovine genome contains almost 3 billion nucleotides with 30 million SNPs. Thus, one SNP is present for every 100 nucleotides in the bovine genome. Each low-cost chip has almost 50,000 genome-wide SNPs for cattle, pig, sheep and chickens. The genomic selection is implemented on selection candidates by measuring the traits and genotyping the markers in a training population. The genomic selection has been adopted extensively for dairy cattle breeding. The genomic estimated breeding values (GEBV) are predicted in selection candidate animals from a prediction model. The genomic selection is based on the LD between SNP and QTL (Fig. 11.4). Since we do not observe the direct association between QTL and the phenotype, the effectiveness of genomic selection is predicted based on how much variance of the trait is explained by the SNP. There are two classes of genomic selection methods: SNP effect-based method and genomic

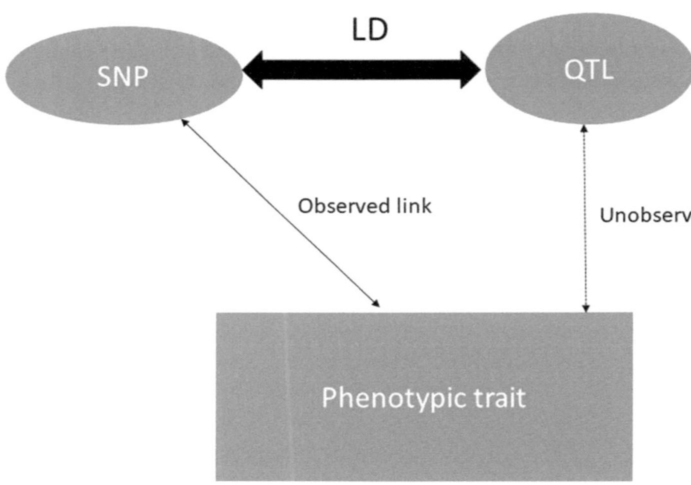

Fig. 11.4 Genomic selection in farm animals is based on linkage disequilibrium (LD) between single-nucleotide polymorphism (SNP) and quantitative trait loci (QTL)

relationship-based method. The RR-BLUP and SNP-BLUP are popular SNP-effect-based methods. Genomic BLUP (GBLUP) is a common genomic-relationship method widely used in animal breeding. The single-step genomic BLUP (ssGBLUP) is another method which combines the phenotype, genotype and pedigree information in single estimate. The accuracy of GEBV has been evaluated in cattle, sheep and pig for many economically useful traits. The reliability of GEBV has a much higher range of 44–49% for milk yield traits in dairy bulls when compared with traditional selection (15–28%). There was also an improvement in accuracies of GEBV for various traits in dairy, meat and wool sheep in comparison to pedigree-based selection. However, accuracy of GEBV for growth and carcass traits improved up to 0.42 and 0.65 respectively in beef cattle. The economic traits like feed conversion ratio were improved significantly in pigs using 60,000 SNPs during genomic selection. The genomic selection is highly useful for those traits expressing after attaining a minimum age for breeding or expressing in one sex. For example, milk yield is a sex-limited trait which only expresses in females.

11.7 Exercises

1. A model simulating the dynamics of lactating mammary glands of cattle is available in the R package ZeBook. This model is explained by six state variables, namely H (changes in level of hormones), CS (production and loss of milk secreting cells), M (removing the secretion of milk), Mmean (the average quantity of milk in the animal), RM (the amount of milk removed) and Y (yield of the milk). Find the fourteen parameters involved in the production process and simulate the model with a time step dt = 0.1 assuming regular consumption of milk by a calf. Find the weekly changes in the state variables of the milk production model.

Solution (Fig. 11.5)

```
> library(ZeBook)
> lactation.define.param(type = "calf")
cu kdiv  kdl  kdh    km ksl    kr  ks ksm mh mm  p mum rc
nominal 1000  0.2 4.43 0.01 0.005   3 0.048 0.1 0.2 27 30 10   1 40
binf      NA   NA   NA   NA    NA  NA    NA  NA  NA NA NA NA  NA NA
bsup      NA   NA   NA   NA    NA  NA    NA  NA  NA NA NA NA  NA NA

>X<-lactation.calf.model2(lactation.define.param()["nominal",],300,0.1)
>X
```

	week	M	Mmoy	CS	RM
[1,]	0	1.959912	0.1979127	2660.532	11.41181
[2,]	1	6.481227	1.3410402	5458.517	23.4589
[3,]	2	11.327486	3.5510203	6806.066	28.68241
[4,]	3	14.63367	6.3156298	7417.833	30.69344
[5,]	4	16.130573	8.9517245	7663.076	31.38036
[6,]	5	16.560699	11.0859653	7724.089	31.55814
[7,]	6	16.457568	12.6450191	7690.09	31.51636
[8,]	7	16.069055	13.6824083	7604.779	31.35529
[9,]	8	15.515307	14.2853523	7491.044	31.11501
[10,]	9	14.867873	14.5422046	7362.316	30.8167
[11,]	10	14.17429	14.532704	7226.456	30.47417
[12,]	11	13.463257	14.3245057	7087.167	30.09551
[13,]	12	12.748816	13.971805	6945.352	29.68349
[14,]	13	12.036879	13.5159428	6800.569	29.23734
[15,]	14	11.331038	12.9876756	6652.105	28.75523
[16,]	15	10.63555	12.4100553	6499.509	28.23602
[17,]	16	9.955806	11.8009379	6342.711	27.68008
[18,]	17	9.29756	11.1746928	6181.953	27.08928
[19,]	18	8.665965	10.5431565	6017.671	26.46661
[20,]	19	8.064959	9.9160827	5850.385	25.81569
[21,]	20	7.497047	9.3013494	5680.637	25.14037
[22,]	21	6.963408	8.7050959	5508.96	24.44449
[23,]	22	6.464142	8.1318754	5335.862	23.73172
[24,]	23	5.998552	7.5848425	5161.833	23.00556
[25,]	24	5.565405	7.0659649	4987.341	22.26928
[26,]	25	5.163134	6.5762413	4812.84	21.52595
[27,]	26	4.789989	6.1159063	4638.772	20.77846
[28,]	27	4.444147	5.6846142	4465.568	20.02954
[29,]	28	4.123776	5.2815939	4293.645	19.28176
[30,]	29	3.827093	4.9057776	4123.404	18.53753
[31,]	30	3.552384	4.5559025	3955.229	17.79909
[32,]	31	3.298024	4.2305899	3789.487	17.06855
[33,]	32	3.062487	3.9284054	3626.518	16.34784
[34,]	33	2.844349	3.6479033	3466.64	15.63874
[35,]	34	2.642285	3.387659	3310.145	14.94287
[36,]	35	2.455066	3.1462907	3157.294	14.26167
[37,]	36	2.281559	2.9224751	3008.32	13.59643
[38,]	37	2.120716	2.7149555	2863.426	12.94826
[39,]	38	1.971572	2.5225479	2722.787	12.31814
[40,]	39	1.833237	2.344142	2586.543	11.70685
[41,]	40	1.704893	2.1787014	2454.81	11.11507
[42,]	41	1.585787	2.0252619	2327.671	10.54328
[43,]	42	1.482007	1.8916573	2212.872	10.02648

Fig. 11.5 Weekly changes in state variables during lactation

2. An R package *agridat* has a dataset of mating crosses of chicken consisting of 45 observations on three variables, namely male parent, female parent and weight of the progeny at 8 weeks. Plot the dataset using R package *lattice*. Estimate the variance component fitting linear mixed model and heritability of both parents as well as combined heritability.

Solution (Fig. 11.6)

```
> library(agridat)
>library(lattice)
> data(becker.chicken)
> chickdata <- becker.chicken
>dotplot(weight ~ female, data=chickdata, group=male,
main="Body weight at 8 weeks", xlab="Female dams",ylab="Progeny
weight",pch=16,auto.key=list(columns=5))
```

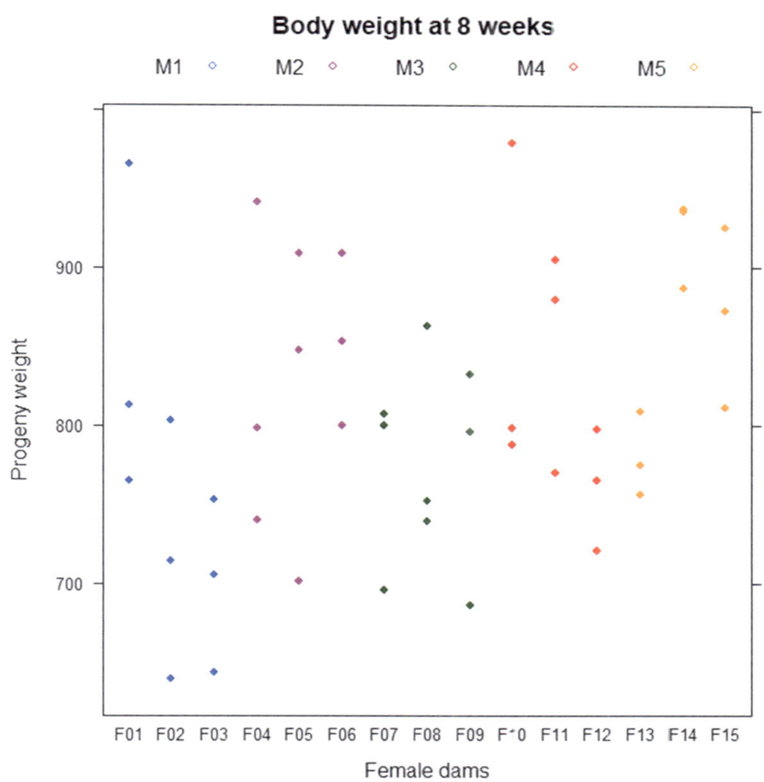

Fig. 11.6 Body weights of chicken at weight weeks linked with male and female parents

```
#Estimation of variance component fitting linear mixed model#
> library(lme4)
> model <- lmer(weight ~ (1|male) + (1|female), data=chickdata) # a model
that incorporates both fixed- and random-effects term
> model
Linear mixed model fit by REML ['lmerMod']
Formula: weight ~ (1 | male) + (1 | female)
Data: chickdata
REML criterion at convergence: 516.6906
Random effects:
 Groups  Name        Std.Dev.
 female  (Intercept) 33.10
 male    (Intercept) 27.87
 Residual            74.33
Number of obs: 45, groups: female, 15; male, 5
Fixed Effects:
(Intercept)
   808.5
#Extract the variance components from a fitted model#
>library(lucid)
> vc(model)
   grp      var1 var2  vcov sdcor
  female (Intercept) <NA> 1096  33.1
   male (Intercept) <NA> 776.8 27.87
 Residual     <NA> <NA> 5524  74.33
> Male.var <- 776
> Female.var <- 1095
> Withincross.var<- 5524
> Vp <- Male.var+ Female.var + Withincross.var
> 4*Male.var/Vp
[1] 0.4197431 # Heritability of male sires
> 4*Female.var/Vp
[1] 0.5922921  # Heritability of female dams
>2*(Male.var+Female.var)/Vp
[1] 0.5060176 # combined heritability
```

11.8 Multiple Choice Questions

1. The farm animal production system consists of:
   (a) Biology of farm animal
   (b) Environment of animal

   (c) Management of the farm
   (d) All the above
2. The modelling of experimental data in farm animal system is known as:
   (a) Mechanistic model
   (b) Empirical model
   (c) Data driven model
   (d) None of the above
3. The adaptation of farm animal to a new nutritional regime is under control of:
   (a) Homeorhetic regulation
   (b) Homeostatic regulation
   (c) Haemorrhagic regulation
   (d) Nutritional regulation
4. The gonadal maturity of a laying hen is primarily determined by:
   (a) Feeding regime
   (b) Genetic merit
   (c) Photoperiod
   (d) Temperature
5. The optimization of laying hen model is performed by:
   (a) Non-linear programming
   (b) Linear programming
   (c) Exponential programming
   (d) Functional programming
6. The stochastic component (s) of a lactation curve is/are:
   (a) Genetic variability among dairy cattle
   (b) Health status of dairy cattle
   (c) Feeding practices
   (d) All the above
7. Linkage disequilibrium measures:
   (a) Random association between the alleles at different loci
   (b) Stochastic association between alleles at different loci
   (c) Non-random association between alleles at different loci
   (d) None of the above
8. A bovine genome contains:
   (a) ten million SNPs
   (b) 20 million SNPs
   (c) 30 million SNPs
   (d) 40 million SNPs
9. The genomic selection method (s) implemented in dairy cattle is/are:
   (a) SNP-BLUP
   (b) RR-BLUP
   (c) G-BLUP
   (d) All the above
10. The phenomics data of a dairy cattle consist of:
   (a) Morphometric measures
   (b) Physiological measures

(c) Behavioural measures
(d) All the above

Answer: 1. d 2. b 3. b 4. c 5. b 6. d 7. c 8. c 9. d 10. d

Summary

- Empirical model is derived from experimental data to infer relationships among different components at a single level.
- Mechanistic model is a process-based model with several individual components and their specific interactions.
- Homeorhetic regulations control the basic life processes associated with reproduction and growth of an animal.
- Homeostatic regulations operate on the adaptive changes in animals under altered nutritional environment.
- A mathematical model can predict food intake of a hen based on its body protein weight and potential daily egg output.
- The shape of a lactation curve is characterized by an ascending phase after parturition reaching its maximum peak followed by a declining slope till the dry-off of the dairy cattle.
- Bovine Genome Database (BGD) is a web-based resource providing access to bovine genome assemblies and annotations.
- Livestock phenomics data can be measured at different levels from molecular level to morphological level.
- The genomic selection is based on the LD between SNP and QTL.
- Genomic BLUP (GBLUP) is a common genomic-relationship method widely used in animal breeding.

Suggested Reading

Khatib H (2015) Molecular and quantitative animal genetics. Wiley-Blackwell, Hoboken

Mondal S, Singh RL (2020) Advances in animal genomics. Academic Press, London

Isik F, Holland J, Maltecca C (2017) Genetic data analysis for plant and animal breeding. Springer, Cham

Mrode RA (2014) Linear models for the prediction of animal breeding values, 3rd edn. CABI Publishing, Wallingford

Computational Bioengineering

12

Learning Objectives
You will be able to understand the following after reading this chapter:

- Application of bioinformatics in bioengineering.
- Control and systems theory in biological systems.
- Metabolic engineering of biological systems.
- Evolutionary engineering of microbes and directed evolution of proteins.
- Principles and methods of synthetic biology.

12.1 Introduction

Cell is a biochemical factory consisting of small molecules involved in the signalling and energy transfer processes. The availability of high-throughput data and the advent of systems approach in biology have paved the way for understanding complex dynamic interactions among molecules in the cell. Thus, engineering approaches are necessary to understand the overall functioning of the cell systems. A cell can be perceived as a dynamic system characterized by a set of state variables. The transcriptome or proteome at a particular time point represents state variables of a cell. Moreover, a cell can be described as an input/output (I/O) system where an input may be a hormone or drug and a secreted metabolite may represent a measurable output. Similar to engineering systems, cell consists of subsystems with different functions such as cell growth, cell division and apoptosis. In addition to being a dynamical system, a cell can also be treated as a controlled environment where all state variables are interconnected to maintain the robustness of the cell. Thus, computational model of the cell using high-throughput data will help in unravelling the control mechanisms in cell regulating its specific functions.

B. K. Tiwary, *Bioinformatics and Computational Biology*,
https://doi.org/10.1007/978-981-16-4241-8_12

12.2 Control and Systems Theory in Biology

In engineering, a control system consists of interconnected components acting in concert to produce a desired behaviour in presence of multiple external perturbations. Majority of biological systems like their engineering counterparts are controlled by closed loop. The control system of human thermoregulation is illustrated in Fig. 12.1 as an example of closed loop. Here, the output value of the system is fed back in the system in order to adjust the input value and is known as *feedback control*. The control system maintains the time dynamics of a particular control variable in a system with a predetermined reference variable in spite of regular external disturbances. A feedback sensor measures the output and subsequently compares this feedback signal with the desired reference value. Finally, the error or deviation obtained between two values is used to compute *control variable*. The feedback signal is subtracted from the input reference value in the negative feedback control. Here, the controller tends to increase the output in case resulting error is positive, whereas it always decreases the output if resulting error is negative. In contract, the feedback signal is added to the input reference value in the positive feedback control. The maintenance of constant body temperature in human is an example of negative feedback control in a biological system. On the other hand, the propagation of action potential in neuron is regulated by the positive feedback control.

Control theory is an integral part of a dynamical system. The interaction of a dynamical system with surrounding environment occurs through two vectors of time-dependent variables: input variables and output variables. The knowledge of an input variable is not enough to describe the output variables in a dynamical system; a minimum state of variables known as state variables needs to be characterized to predict the future output of a dynamical system. Thus, the knowledge of both input variables and state variables is necessary to predict the overall output of a system. Most of the biological systems like artificial systems are non-linear dynamical systems. Dynamical systems are modelled computationally

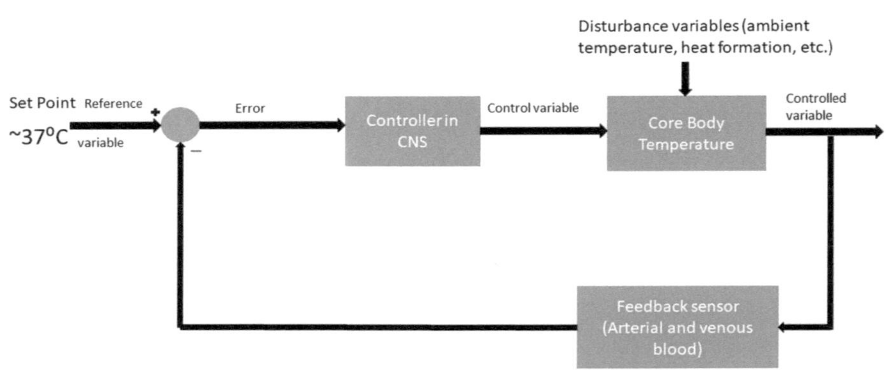

Fig. 12.1 The control system of human thermoregulation using closed-loop feedback

either using two deterministic approaches: continuous time series and integer time points. State variables represent the real value of the time variable (t) in a continuous-time system, whereas integer time points (i.e. t = 1, 2, 3. .) define a discrete-time system. However, deterministic models, although useful, often fail to capture the inherent complexity of a biological system. Therefore, stochastic models like hidden Markov models are developed to represent a biological system.

12.3 Strategies in Bioengineering

There is a huge demand for bio-based products such as biofuels, solvents, organic acids, polymers, and food supplements across the world. In order to produce them in large quantity at low cost, three performance metrics of industrial production, namely titre (the final concentration of the product after bioprocess), yield and productivity, should be very high. The minimum target for titre is about 100 grams per litre in most cases but a higher target of more than 200 grams per litre is feasible in some cases. The yield of a product is defined as mole or gram of the product obtained from mole or gram of the consumed substrate. On the other hand, productivity is defined in two forms: specific productivity and volumetric productivity. The specific productivity is the amount of the product produced in terms of mole or gram per cell per unit time, whereas the volumetric productivity is the amount of product produced per volume per unit time. The importance of any parameter in decision-making depends on the nature of bio-product and the bioprocess. A high titre is not only a useful parameter for cost-effectiveness, it also facilitates separation and purification of the product. A higher yield of a product is vital for production of bulk chemicals as the major cost is incurred in procuring carbon substrates. Productivity covers the overall production cost incurred in a bioprocess such as fermenter size and equipment depreciation cost per year.

12.4 Metabolic Engineering

Biological systems produce a wide range of natural products from precursor molecules available in the surrounding environment. These natural products can be synthesized in large quantities by culturing biological organisms in artificial culture media. However, in majority of cases, the actual yield of the metabolic product is very low resulting in high cost of industrial production. Thus, the primary goal of metabolic engineering is to reduce this cost substantially by producing large quantities of useful metabolites such as pharmaceuticals, enzymes and antibodies. A biological system is akin to a complex circuit composed of specific interactions among various biomolecules such as DNA, RNA, proteins and metabolites. The common computational software for metabolic engineering are listed in Table 12.1.

The modularity in a biological system not only prevails at the circuit level but manifests at each level of biological complexity. Biological pathways can be separated into a series of different parts coordinating a particular biological function.

Table 12.1 Common computational tools in metabolic engineering

Computational tool	Description
BNICE	Discovery of novel metabolic pathways based on generalized reaction rules
RetroPath	A tool for pathway prediction based on the generalized reaction rules
CAMEO	A tool for in silico modelling including algorithms for detecting gene knockout and overexpression target

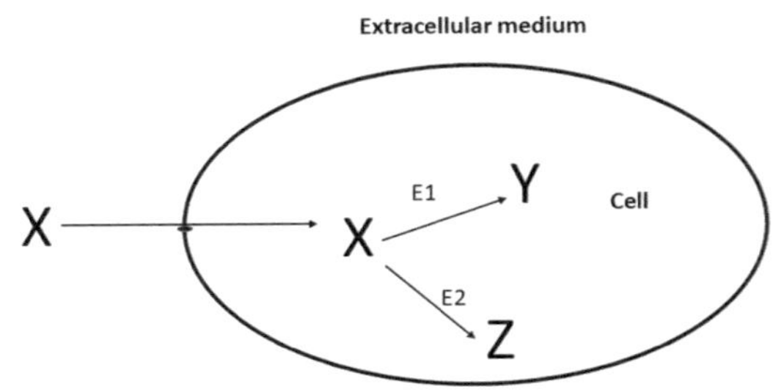

Fig. 12.2 A simple representation of a metabolic network. X is a precursor molecule transported inside a cell via active transport. X is converted to Y and Z through enzymatic reaction catalysed by E1 and E2, respectively

A simple biological module can be represented as promoter with its expressed gene and regulatory proteins with its DNA binding sites. These simple modules in turn give rise to a gene regulatory network which produces an output signal after complex computation in a biological organism when challenged by various inputs. This computation is based on various engineering principles such as positive feedback, negative feedback and feed-forward loop.

The performance of a biological system can be predicted in silico using computational modelling. First, the possible enzyme pathways involved in the biosynthesis of a desired metabolite are identified. A set of reactions in individual paths are stoichiometrically balanced and added to get the net chemical reaction for a path. A metabolic engineer redesigns the existing network pathways in order to change the metabolic flux rates. If one wishes to increase the production of a particular metabolite, an attempt will be made to increase the flux of the metabolite from its precursor without affecting the remainder of the metabolic pathways. The membrane flux of a precursor molecule in a cell is increased by increasing the concentration of that molecule in culture media in case of passive diffusion. In contrast, the concentration or efficiency of a transporter protein is increased for active transport of precursor across the cell membrane. The precursor molecule in the cell is converted to a desired metabolite through multi-step enzymatic reactions. For an enzymatic reaction, if substrate concentration is much lower than the Michaelis constant, the reaction flux will be proportional to the substrate concentration. For example, consider an example where a precursor molecule X enters a cell through transport process and is converted to either Y or Z inside a cell via enzymatic reactions as illustrated in Fig. 12.2. If we want to overproduce metabolite Y, the mass flux from X to Y needs to be increased through enhanced activity of enzyme catalysing this reaction. This enhanced enzyme activity can be achieved by alteration of specific activity of the enzyme (k_{cat} / K_m) or alteration of the concentration of the intracellular enzyme. However, this kind of engineering approach is too simplistic to be accepted by nature. The modified cell is shifted away from optimal state achieved through evolution and shows adverse effects of modifications in form of stunted growth and reduced robustness. There is no specific single rate limiting step in a metabolic network, therefore, perturbations are needed at multiple steps to achieve the desired output of a metabolite. This kind of engineered changes generally are not accepted by the regulatory network of a cell. Sometimes, the enzyme synthesis/activity is repressed by the product metabolite; otherwise, the product metabolite potentiates the competing enzyme activity. Therefore, the best strategy for metabolic engineering is to perform minimum changes in the overall metabolic network of the cell so that the resulting cell is akin to the unengineered system already optimized for growth and reproduction under evolutionary selective processes. Typically, only fluxes to a desired metabolite product are changed keeping the concentrations of other intermediate metabolites unchanged. However, this kind of differential flux balance is difficult to achieve in practice.

The metabolic systems are used as constraints to optimize an objective function, for example, how to maximize the biomass of cells? The prediction of flux balance under a primary objective function will predict the changes in a secondary objective such as metabolite overproduction after perturbations in the metabolic network. The genetic changes in the network such as addition of a gene or deletion of a gene can be tested computationally by suitable modifications in the stoichiometric matrix. The optimal fluxes are recomputed in order to achieve a greater flux towards a desired metabolite. Here, absolute growth rate may not be appropriate objective function; instead, minimal flux disturbance criterion performs better as an alternative objective.

12.5 Evolutionary Engineering

Evolutionary engineering is a discipline of engineering to develop phenotypes of microbial strains or a biological component like a protein molecule simulating natural evolutionary processes. Evolutionary engineering operating on the cells for improvement of strains is known as adaptive laboratory evolution (ALE). When evolutionary engineering is applied on a protein molecule for better efficiency such as enhanced catalytic activity, stability or tolerance to high temperature, it is known as directed evolution. In both processes, first variants are randomly generated using natural evolution and then desired protein or cells are selected from large number of naturally derived variants. In ALE, microbial cells are cultured in defined culture media under controlled conditions for long period of time (100–2000 generations) in order to achieve improved phenotypes. Most common mutations observed during adaptive laboratory evolution are single-nucleotide polymorphisms (SNPs), small insertions and deletions (indels), transposition and deletion of large genomic regions contributing towards genetic and gene regulatory changes in the improved strains. The increased fitness of improved strain in culture is manifested by increasing frequency of this strain among total population. Microbial cells are exposed to various nutrients such as glycerol and glucose in chemostat cultures for improved growth. Similarly, the effects of environmental stresses such as high temperature and osmotic pressure or tolerance to high ethanol concentration on the adaptive laboratory evolution of microbial cells have tremendous industrial applications. Computational models are useful in understanding the evolution of laboratory strains. However, the major challenges are the high computational cost of simulation and unknown functions of a significant proportions of genes (~30%) in a model species like *E. coli* or *S. cerevisiae*. The effects of various environmental stressors on microbial cell are also very difficult to integrate in the model. In spite of limitations, computational models can play a significant role in improving laboratory strains through adaptive laboratory evolution.

The directed evolution has been applied on the artificial evolution of enzymes to catalyse the industrial reactions. Frances Arnold received the Nobel prize in chemistry in 2018 for her pioneering work on the directed evolution. Darwinian evolution favours the fit organisms by accumulating beneficial mutations. In directed evolution, macromolecules are purposefully evolved towards new desired properties. The evolutionary processes are implemented in three successive steps in directed evolution: mutation, selection and amplification. First, a natural macromolecule whose properties are similar to the desired one is taken as a precursor molecule for redesigning. Alternatively, a new macromolecule is created de novo from a collection of random sequences and the process is known as protein design. During redesigning, catalytic activity of an enzyme can be transplanted on another enzyme with same folds and mechanistic functions. Even a few mutations at the active site of the enzyme facilitates emergence of novel functions. Computational redesign of an extant protein is generally associated with the experimental evolution of the protein. For example, the computational design starts with simulated docking of a target ligand into the active site of a protein retrieved from the protein data bank (PDB). The residues around this pocket are randomized and the design algorithm repeatedly searches for a conformational space of the side chains and ligands at the minimum energy level for each sequence. Finally, only a few proteins are experimentally produced from a list of sequences with minimum energy predicted by the computational design algorithm. Novel receptors for proteins and small molecules have been designed using this approach. A variety of experimental methods such as error-prone PCR, degenerate codons, DNA shuffling and recombination are used in vitro to create diversity in the molecule. Directed evolution is an established method for enzyme engineering. The specificity, stability and reaction conditions of many enzymes have been custom-made for commercial exploitation. Lipases, generated through directed evolution, are produced commercially as an additive to laundry detergents to break down lipids. An enzyme becomes several hundred or thousand times more effective than the template enzyme after a few iterative cycles of directed evolution.

12.6 Synthetic Biology

Synthetic biology is an emerging discipline integrating engineering and computational biology. It mainly deals with either construction of new artificial system or redesigning a natural biological system. The basic principle of this discipline is to simplify a complex biological system into much simpler molecular factories consisting of living parts and devices to produce a biological product. Biological circuits are assembled from the genes and regulatory DNA (similar to gates in electronic circuits) and the proteins are akin to the wires in the electronic circuits. The parts, devices and systems are defined in the framework of synthetic biology which is as follows:

Parts The parts or bioparts is a portion of DNA molecule encoding a specific biological function. For example, promoter, ribosome binding site, protein coding sequence and terminator sequence are four separate parts of a genetic circuit producing a polypeptide. The standardized and interchangeable biological parts are available in the Registry of Standard Biological Parts and Biobricks.

Devices Devices are made of different parts performing a user-defined function. For example, a simple device protein generator consists of promoter, ribosome binding site, protein coding sequence and terminator sequence. The devices have logic gates having binary states 1 (condition satisfied) and 0 (condition not satisfied). A logic gate takes multiple binary inputs and produces a single output in form of either 0 (OFF) or 1 (ON). There are different types of logic gates in a device such as AND gate (all input conditions must be satisfied for ON state), OR gate (any input condition must be satisfied for ON state), NOT gate (single input is inverted with compatible promoter/repressor pair) and negated gate (a combination of NOT gate with AND/OR gate).

Systems Simple systems are developed from one or more devices. The examples of simple biological systems are feedback loops, genetic toggle switches, oscillators and repressilators. The effect of output on the future behaviour of a system is known as feedback effect. When the future output of a system is increased by the present output, it is known as positive feedback. Conversely, when a present output decreases the future output of the system, it is known as negative feedback. Activators and repressors are used for creating positive and negative feedback respectively in synthetic biology. The switch is a device which can be turned ON or OFF in response to an input signal. Toggle switches are having only two stable steady states ON and OFF without any intermediate state. Toggle switches in synthetic biology are designed by integrating two repressors and two constitutive promoters, where each promoter transcribe a repressor which in turn inhibit the opposite promoter. An oscillator is a device showing cyclical pattern (oscillations) in the output nearby an unstable steady state. There are two important elements in a biological oscillator: an inhibitory feedback loop and a source of delay in the feedback loop. A repressilator is a kind of oscillator made by combining multiple inverters. For example, three genes may be combined in feedback loop in such a way that each gene represses the subsequent gene in the loop and itself repressed by the preceding gene.

The synthetic organisms assembled from genetic parts and devices do not always exhibit predictable behaviour unlike its electronic counterparts due to stochastic noise and uncertain environment in a new chassis. Sometimes, engineered cells show low productivity due to metabolic burden that is the part of the cellular resources used in engineering the metabolic pathways. The user-defined function such as synthesis of a new product in an engineered cell may jeopardize the overall metabolic robustness optimized by natural selection in an evolutionary timescale.

12.7 Computational Design of Synthetic Systems

The rapid development of synthetic biology is a result of collaborative efforts of both computational and experimental biologists. The rational design of a synthetic system needs a prior design and optimization of the computational model in standard system before translating or compiling into DNA sequences in vitro or in vivo. Synthetic Biology Open Language (SBOL) was developed as a standard format to support the specification and exchange of information regarding biological design. Systems Biology Markup language (SBML) can only represent a model of a biological process but lacks information regarding associated nucleic acids and amino acid sequences. SBOL framework have a modular and hierarchical graph based structural and functional description of genetic design components, their sequences and inter-component interactions. A design repository, SynBioHub, is an open-source community-based effort to share information regarding synthetic systems. It uses SBOL as a core storage data format and allows a user to upload, browse and share designs on the web interface. There is a harmonized data exchange between multiple existing SynBioHub repositories. We can transform the textual information of a biological design to visualization of the design using the web interface of a software tools like Pigeon and TinkerCell. There are many computer-aided circuit design applications for synthetic biology (Table 12.2). BioJADE was the first drag and drop JAVA-based tool for circuit design which allows use of parts from BioBrick repositories. The list of parts is visible on the workspace and they are connected through hypothetical wires. Unlike electronic circuits, biological parts and devices are

Table 12.2 Computational tools for circuit design

Tool	URL
BioJADE	http://web.mit.edu/jagoler/www/biojade/
TinkerCell	http://www.tinkercell.com/
ProMoT	https://www2.mpi-magdeburg.mpg.de/projects/promot
GenoCAD	https://design.genofab.com/
GEC	https://omictools.com/gec-tool
Cello	www.cellocad.org

connected through a variety of signalling molecules. For example, RNA polymerases and ribosomes act as common signalling carriers for exchanging information between the parts. Thus, the biological current between the parts are represented by their fluxes and measured in terms of Polymerase per second (PoPS) and ribosomes per second (RiPS). BioJADE and TinkerCell consider only polymerase as a carrier of signal between the parts. TinkerCell is capable of running both stochastic and deterministic simulations in addition to other analyses such as flux balance and sensitivity analysis. Asmparts is another command line tool which treats each part as independent SBML module and can run model kinetics of transcription and translation. Process Modelling Tool (ProMoT) is a graphical tool for process engineering which has additional signal carriers like transcriptional factors, chemicals and small RNAs in addition to polymerases and ribosomes. It is used for modelling both kinetic models and Boolean models and resulted models can be exported to SBML and MATLAB format. GenoCAD and GEC are two other tools connected with databases for circuit design.

The hardware description language like Verilog allows an electronics engineer to design an integrated circuit using textual commands. This computational design is subsequently transformed into silicon circuits in an electronic system. This concept is recently translated to genetic circuits where a Verilog design was performed in a computational environment known as Cellular Logic (Cello) and subsequently translated to a linear DNA sequence. Cello builds genetic circuits using transcriptional gates with RNA polymerase as signal carrier and also simulates the performance of the genetic circuit.

The designed circuits are optimized to evaluate the performance efficiency. Two classes of optimization methods have been applied till date in synthetic biology: stochastic and deterministic. Stochastic optimization process such as evolutionary algorithm is computationally expensive and finds solution through random search. In fact, it often fails to find global optimal solution. In contrast, deterministic methods such as gradient descent have local search algorithms with lower computational cost. It often fails to get many better solutions during the optimization process. Evolutionary algorithm requires the initial definition of basic components, biochemical reactions, ensemble of independent networks and objective function. The best networks are obtained as a final solution after several rounds of iterations. Genetdes is a stochastic method for finding a single solution from random circuit configuration but often get stuck in local minima. Conversely, OptCircuit is a deterministic method for finding a local optimal solution very close to the best one. However, these methods can only be implemented on simplified models of small genetic circuits. RoVerGeNe is a software specially designed to adjust the performance of a synthetic network and to measure its robustness.

Box 12.1 Risks Associated with Synthetic Biology

A synthetic microbe, algae, plants or animals may become pathogenic to other organisms in the ecosystem and degrade the natural environment. Further, it may outcompete with wild native populations and adapt even in an adverse environment. The modified algal strain may cause algal blooms that creates harmful effect on human health. The ecological impact of intentional release or accidental escape of a synthetic organism in the environment is a decline in biodiversity and genomic diversity of a species. The horizontal gene transfer of engineered genes from modified organisms to wild-type organisms may have catastrophic consequences. The designer protein toxins developed using protein engineering for therapeutic purpose may be used as a bioweapon for military purpose or bioterrorism. A synthetic organism may act as an allergen or generate an allergen protein to human. The products generated during biofuel production may be carcinogenic to human and other animals.

Box 12.2 iGEM

The International Genetically Engineered Machine (iGEM) is a competition for high-school, undergraduate and postgraduate students to develop a project in synthetic biology. This program was started in 2005 by the MIT. Each team consists of 5–15 students under supervision of a qualified researcher. The newly registered team gets a set of thousand standard DNA parts stored in the BioBrick to start their project. The best new synthetic biology ideas in different category are rewarded in this competition. Many projects of the iGEM competition have been published in reputed international journals like Nature, Science and Cell. Some of the noted examples of the iGEM project are bacterial photofilm producing light-induced pigments and different smells produced by bacteria during exponential phase and stationary phase of their growth.

Box 12.3 Engineered Riboswitches

Riboswitches are natural regulatory *cis*-acting RNAs that bind small molecules in order to perform their regulatory functions on the target genes. They have specific ligand binding pockets containing aptamer domain sites in the mRNA. The selective binding of a ligand at an aptamer site initiates a conformational change in the RNA molecule leading to altered gene expression. Synthetic transacting riboregulator tools known as antiswitches have

(continued)

Box 12.3 (continued)
been created by fusion of antisense-effector domains with riboswitch-like aptamer domains in yeast. The binding of a small-molecule theophylline to the aptamer domain leads to a conformational change in the synthetic molecule allowing antisense domain to bind the target mRNA including reporter green fluorescent protein (*gfp*)gene encoding the GFP protein. Moreover, the thermodynamic properties of antiswitches and their overall system response can be changed by introducing targeted mutations. These antiswitches including various combination of aptamer and antisense domains have a tremendous potential in the post-transcriptional modulation of gene expression in eukaryotes. They provide us a novel synthetic tool to activate or repress any target gene in response to a small molecule.

12.8 Exercises

1. An R package BiGGR contains the BIGG model (iND750) of a yeast cell. Yeast performs alcoholic fermentation using glycolysis to produce pyruvic acid from glucose followed by production of ethanol from pyruvic acid. Plot the metabolites and reactions involved in glycolysis in the yeast model using BiGGR package.

Solution (Fig. 12.3)

```
library(BiGGR)
> data(S.cerevisiae_iND750)
> model <- S.cerevisiae_iND750@model
> relevant.species <- c("M_glc_DASH_D_c", "M_g6p_c", "M_f6p_c",
+ "M_fdp_c", "M_dhap_c", "M_g3p_c",
+ "M_13dpg_c", "M_3pg_c", "M_2pg_c",
+ "M_pep_c", "M_pyr_c",
+ "M_6pgl_c", "M_6pgc_c", "M_ru5p_DASH_D_c",
+ "M_xu5p_DASH_D_c", "M_r5p_c", "M_g3p_c", "M_s7p_c")
> relevant.reactions <- c("R_HEX1", "R_PGI", "R_PFK", "R_FBA",
"R_TPI",
+ "R_GAPD", "R_PGK", "R_PGM", "R_ENO", "R_PYK",
+ "R_G6PDH2r", "R_PGL", "R_GND", "R_RPE", "R_RPI", "R_TKT1")
> hd <- sbml2hyperdraw(model,
+ relevant.species=relevant.species,
+ relevant.reactions=relevant.reactions,
+ layoutType="dot", plt.margins=c(20, 0, 20, 80))
> plot(hd)
```

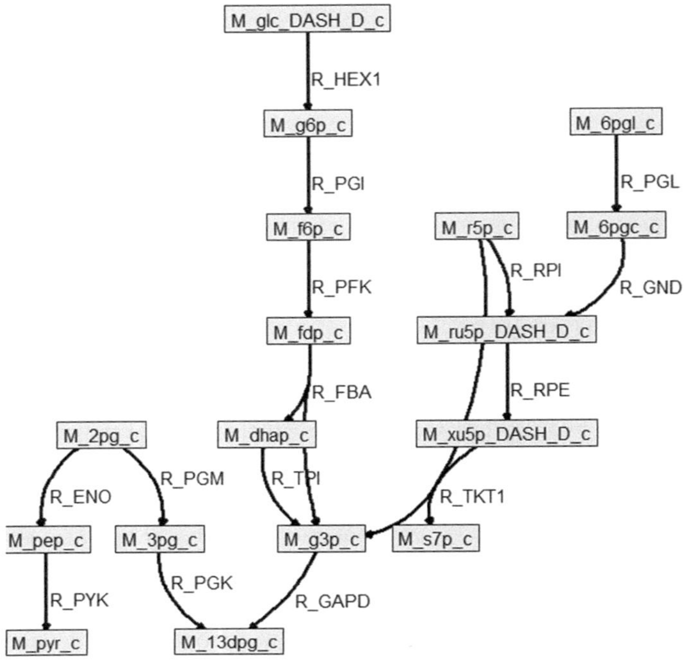

Fig. 12.3 Glycolytic metabolites and reactions in the yeast model

2. Fluxer (https://fluxer.umbc.edu) is a web application for computation of genomic-scale flux networks . It can compute and visualize the k-shortest path between metabolites or reactions in order to find main metabolic route. Using this web application, perform the following tasks on the yeast model (iND750):
   (a) Visualize the complete yeast model in force-based layout.
   (b) Find the k-shortest path between extracellular glucose and intracellular glucose along with flux.
   (c) Find the k-shortest path from extracellular glucose to the objective function of maximum biomass. Knock out the glucose transport uniport and observe the effect of this knockout on this metabolic pathway.

Solution

(a) The complete graph of genome-scale metabolic model of yeast is as follows (Fig. 12.4):
(b) The k-shortest path between extracellular glucose and intracellular glucose is as follows. The flux of this exchange reaction is 1 unit (Fig. 12.5).
(c) The k-shortest path from extracellular glucose to the objective function of maximum biomass along with fluxes is as follows (Fig. 12.6):

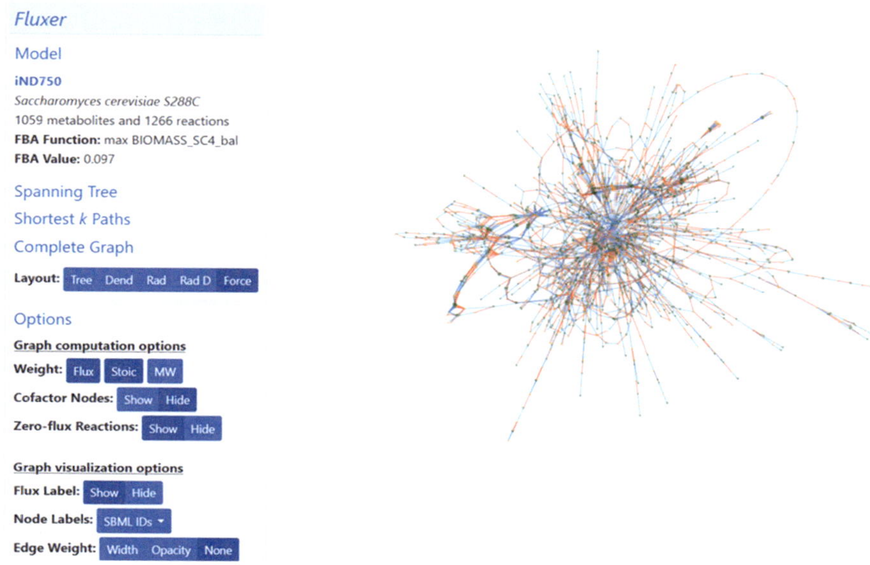

Fig. 12.4 The complete graph of a genome-scale metabolic model of yeast

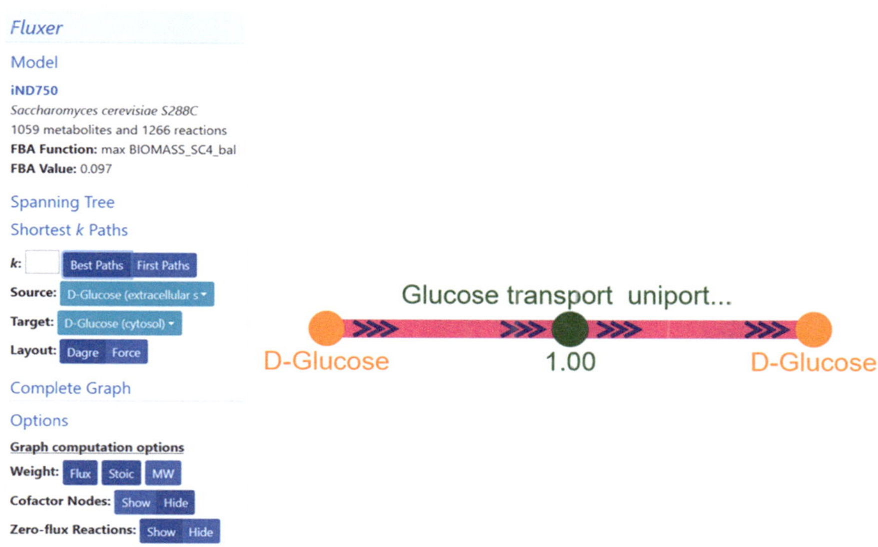

Fig. 12.5 The k-shortest path between extracellular and intracellular glucose

Fig. 12.6 The k-shortest path from extracellular glucose to maximum biomass of yeast along with fluxes

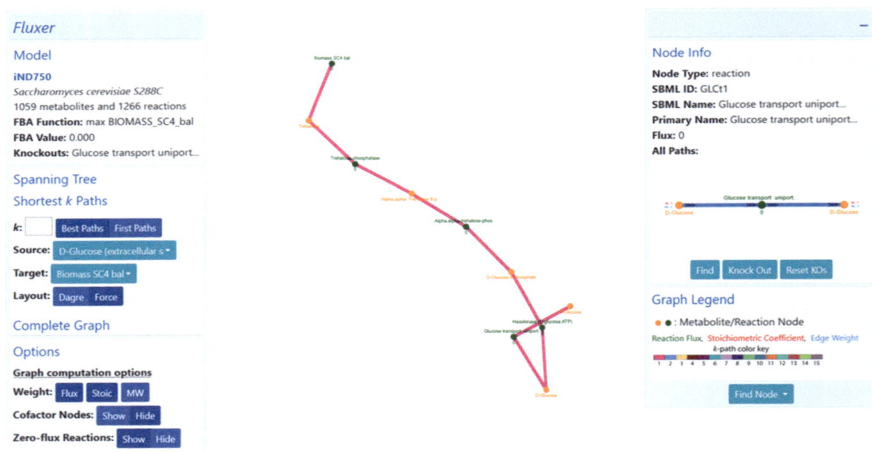

Fig. 12.7 The k-shortest path from extracellular glucose to maximum biomass of yeast showing zero fluxes after knocking down glucose transport uniport

The knockout effect of glucose transport uniport on the metabolic pathway is as follows. Interestingly, all reactions are having zero flux due to non-availability of glucose as a sole source of energy (Fig. 12.7).

12.9 Multiple Choice Questions

1. Propagation of action potential is an example of:
   (a) Negative feedback control
   (b) Positive feedback control
   (c) Positive feed-forward control
   (d) None of the above
2. Majority of biological systems are:
   (a) Non-linear dynamical systems
   (b) Linear dynamical systems
   (c) Linear static systems
   (d) None of the above
3. The stochastic models of biological systems are represented by:
   (a) Hidden Markov models
   (b) Continuous time series models
   (c) Integer time points models
   (d) None of the above
4. An improvement of laboratory microbial strains is achieved through:
   (a) Continuous laboratory evolution
   (b) Random laboratory evolution
   (c) Adaptive laboratory evolution
   (d) Stochastic laboratory evolution
5. Directed evolution of proteins is performed using:
   (a) Error-prone PCR
   (b) Degenerate codons
   (c) DNA shuffling
   (d) All the above
6. Toggle switches in synthetic biology consist of:
   (a) Two repressors and two constitutive promoters
   (b) Two activators and two constitutive promoters
   (c) One activator and one constitutive promoter
   (d) One repressor and one constitutive promoter
7. A repressilator is an oscillator made of:
   (a) A combination of multiple inverters
   (b) A combination of multiple activators
   (c) A combination of multiple promoters
   (d) A combination of multiple repressors
8. The standard format for specification of biological design is known as:
   (a) SBML
   (b) SBOL
   (c) SBAL
   (d) SBCL

9. The JAVA-based tool for biological circuit design is known as:
   (a) BioJAVA
   (b) BioJADE
   (c) BioCAD
   (d) None of the above
10. The deterministic method to find a local optimal solution is known as:
   (a) OptCircuit
   (b) OptSol
   (c) OptSolution
   (d) OptOutput

Answer: 1. b 2. a 3. a 4. c 5. d 6. a 7. a 8. b 9. b 10. a

Summary

- A cell can be perceived as a dynamic system characterized by a set of state variables.
- Both biological systems and artificial systems are non-linear dynamical systems.
- A metabolic engineer redesigns the existing network pathways in order to change the metabolic flux rates.
- Best strategy for metabolic engineering is to perform minimum changes in the overall metabolic network of the cell.
- Computational models are useful in understanding the evolution of laboratory strains.
- Directed evolution has been applied on the artificial evolution of industrial enzymes.
- Synthetic biology deals with either construction of new artificial system or redesigning a natural biological system.
- Common examples of simple biological systems are feedback loops, genetic toggle switches, oscillators and repressilators.
- Toggle switches have only two stable steady states ON and OFF without any intermediate state.
- Synthetic Biology Open Language (SBOL) is a standard format to support the specification and exchange of information regarding biological design.

Suggested Reading

Filipovic N (2020) Computational bioengineering and bioinformatics: computer modelling in bioengineering. Springer, Cham

Filipovic N (2019) Computational modeling in bioengineering and bioinformatics. Academic Press, London

Zhang G (2017) Computational bioengineering. CRC Press, Boca Raton

Smolke C (2018) Synthetic biology: parts, devices and applications. Wiley-Blackwell, Weinheim

MIX
Papier aus verantwortungsvollen Quellen
Paper from responsible sources
FSC® C105338

If you have any concerns about our products,
you can contact us on
ProductSafety@springernature.com

In case Publisher is established outside the EU,
the EU authorized representative is:
Springer Nature Customer Service Center GmbH
Europaplatz 3, 69115 Heidelberg, Germany

Printed by Libri Plureos GmbH
in Hamburg, Germany